Geological Pioneers of the Jurassic Coast

Geological Pioneers of the Jurassic Coast

ANDREW S. GOUDIE AND DENYS BRUNSDEN

Oxford University Press is a department of the University of Oxford. It furthers the University's objective of excellence in research, scholarship, and education by publishing worldwide. Oxford is a registered trade mark of Oxford University Press in the UK and certain other countries.

Published in the United States of America by Oxford University Press
198 Madison Avenue, New York, NY 10016, United States of America.

CIP data is on file at the Library of Congress

ISBN 978-0-19-763808-8

DOI: 10.1093/oso/9780197638088.001.0001

1 3 5 7 9 8 6 4 2

Printed by Marquis, Canada

Contents

Acknowledgements

We would like to acknowledge the help given, at many stages of this project, by friends and associates who have provided support, encouragement, and information. In the early stages of the development of the World Heritage idea, David Andrews of Devon County Council, Andy Price, Malcolm Turnbull, Tim Badman, and Richard Edmonds worked tirelessly to prepare the Nomination Document and the supporting scientific, planning, and management files.

Some of the material from these papers is incorporated into this volume. One of the inspirations for the content was *The Heyday of Natural History, 1820–1870*, by Lynn Barber, published by Jonathan Cape in 1980. This is highly recommended for anyone interested in the role of women and the impact of 'outdoor studies' on the general public of the time.

The first drafts of the book were greatly improved by the staff and investments of the Jurassic Coast Trust. We particularly thank Luck Culkin, Anjana Khatwa, and Sam Scriven for their help. Above all we thank Sybil King, former Trustee, and the Fines Foundation, for sending our manuscript to the publisher. She has been a strength to us throughout, and we would not have made progress without her. She should be given thanks from all of us for incredible support to the Site over many years and to many of the learning centres along the coast. We also acknowledge the constant support of our wives, Heather and Elizabeth.

Except where stated otherwise, we believe that the images we have used are in the public domain, and the sources we have used have been given.

The Authors

Andrew S. Goudie DSc, Emeritus Professor of Geography and former Pro-Vice-Chancellor at the University of Oxford, Honorary Fellow of Hertford College, and the former Master of St Cross College, is a recipient of a Royal Medal from the Royal Geographical Society, the Mungo Park Medal of the Royal Scottish Geographical Society, and the Farouk El-Baz Award of the Geological Society of America. He has been Chair of the British Geomorphological Research Group, President of the Geographical Association, and President of the International Association of Geomorphologists. He is the author of *Discovering Landscape in England and Wales* (1985), *The Landforms of England and Wales* (1990), *Great Warm Deserts of the World* (2002), *The Human Impact* (eighth edition, 2018), *Great Desert Explorers* (2016), and *Landscapes and Landforms of England and Wales* (2020). He is a former Trustee of the Jurassic Coast Trust.

Denys Brunsden OBE, DSc, FKC, Emeritus Professor, King's College, London, is a geomorphologist specialising in landslides and coastal erosion, the founder of the Jurassic Coast World Heritage Site, former Chairman of the British Society of Geomorphology, Geographical Association, and International Association of Geomorphologists. He was also the first Chairman of the Dorset Coast Forum. He proposed the Dorset and East Devon Coast for World Heritage Site designation and worked with Malcolm Turnbull, Tim Badman, and many others to achieve this and to write the scientific case. He co-authored with Tim Badman *The Official Guide to the Jurassic Coast, Dorset and East Devon's World Heritage Coast: A Walk Through Time* (2003). In 2010, Denys was awarded the R. H. Worth Prize of the Geological Society for this work and had previously received the William Smith and Glossop Medal awards from the Society. Denys was a Trustee of the Jurassic Coast Trust until 2017, when he became a Patron.

1
Introduction

> This area is truly world-famous for its many and varied geological and palaeontological resources. These resources have played a fundamental role, historically, in the development of basic concepts of earth history and in documenting past life. The region under consideration has been under continuous investigation for more than two centuries. There remains still more, much more, that the . . . Dorset and East Devon Coast has to offer Earth scientists. . . . The more an area is studied and published on, the more it becomes a reference to future generations of investigators.
>
> —Professor Zofia Kielan-Jaworowska (1925–2015),
> Polish Academy of Sciences

The purpose of this book is to recount the lives and achievements of some of the great geologists and geomorphologists who have studied the Jurassic Coast, England's only natural World Heritage Site, and have made it the Mecca that it has become for all those interested in Earth Science since the late seventeenth century. We classify these individuals into six groups: the earliest investigators, the fossil collectors, the geologists of the so-called Golden Age, the geological mappers, gifted stratigraphers and palaeontologists, and remarkable amateurs and some other stars. This is not a perfect classification, and, for example, Marie Stopes, while remarkable, was perhaps as much a professional as an amateur.

The Jurassic Coast includes approximately 155 kilometres of coastline between Orcombe Rocks, at the mouth of the River Exe in Devon, and the geological boundary between the Cretaceous and later strata in Studland Bay in east Dorset (Figure 1.1). It embraces the towns of Budleigh Salterton, Sidmouth, Seaton, Lyme Regis, Charmouth, West Bay, Portland, Weymouth, and Swanage. The Dorset and East Devon Coast was inscribed on the World Heritage List in Helsinki on 13 December 2001. The site was granted World

Geological Pioneers of the Jurassic Coast. Andrew S. Goudie and Denys Brunsden, Oxford University Press.
 DOI: 10.1093/oso/9780197638088.003.0001

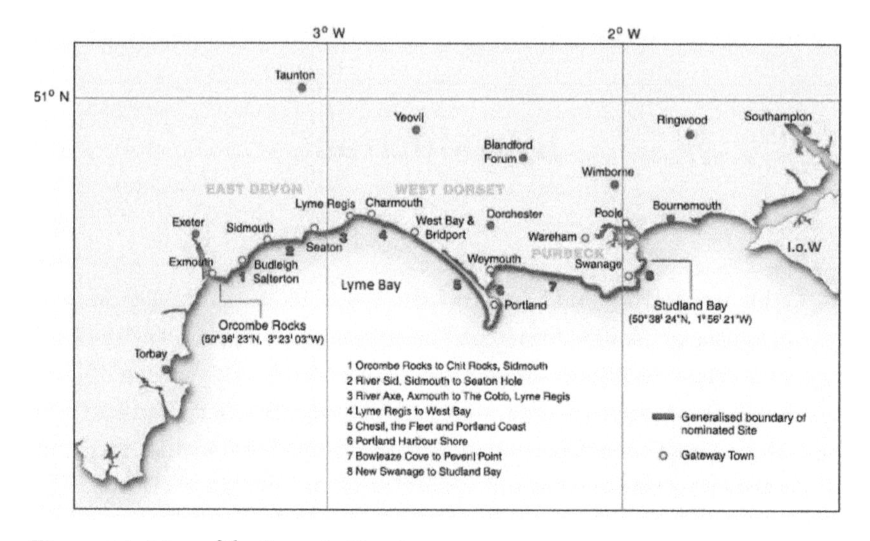

Figure 1.1 Map of the Jurassic Coast.
From © Dorset County Council, Nomination Document, figure 2. With permission of the Jurassic Coast Trust.

Heritage status under UNESCO's criterion viii—Earth's history and geological features—which indicated that its geology and geomorphology were of Outstanding Universal Value. It is currently managed by the Jurassic Coast Trust, an independent charity.

The Geological Backdrop

In this brief section, before moving on to our consideration of the great individuals who have studied the Jurassic Coast, we provide a short introduction to the geology and geomorphology of the area. This provides a necessary backdrop within which the work of these scientists can be placed.

East Devon and Dorset have been known since the early days of geology as providing one of the world's finest sequences of strata from the Triassic of East Devon to the Cretaceous rocks in the East of Dorset (Figure 1.2). They range from the red Triassic rocks of East Devon between Exmouth and Seaton, to the dark grey Lower Jurassic Lias, just west of Lyme Regis, to the yellowish sandstones of West Bay (Figure 1.3), to the great white limestones of Portland, and to a near complete sequence of Cretaceous rocks from the headland of White Nothe (Figure 1.4), through to Old Harry Rocks near

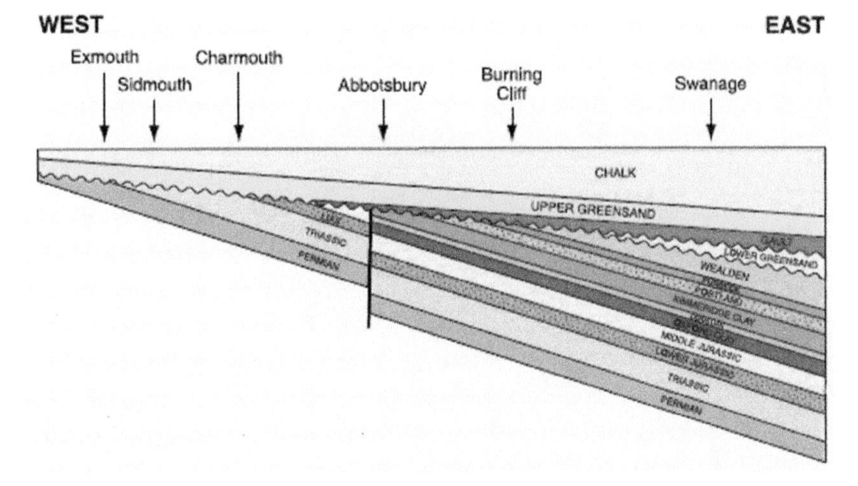

Figure 1.2 Geological cross-section of the Jurassic Coast from West to East (from © Dorset County Council, 2000, Nomination Document, figure 5). In general, the strata dip gently to the east. The oldest rocks are therefore found in the west, with progressively younger strata outcropping to the east. As a direct result, most parts of the succession are readily accessible in sequential order within the cliffs and foreshore.

With permission of the Jurassic Coast Trust.

Figure 1.3 The yellowish early Jurassic Bridport Sand cliffs of West Bay (ASG).

Figure 1.4 The chalk cliffs east of White Nothe (ASG).

Swanage. Further to the east, just outside the area of the Jurassic Coast, there are also Paleogene beds, laid down between 66 and 23 million years ago. Hart's book (2009) provides a beautifully illustrated guide to the geology and landscapes of the region (also see Box 1.1).

The coastline's rocks are displayed in a magnificent series of cliff exposures. Along with such places as Whitby in Yorkshire and Solnhofen in Germany, the area has played a fundamental role in understanding the palaeontology of this period of geological time.

Box 1.1 Some Key Literature

The *Special Memoir of the British Geological Survey* (Barton et al. 2011) gives a full account of the whole coast's geology, as does John Cope's (2016) *Geology of the Dorset Coast*. Ian West's remarkable website, *The Geology of the Wessex Coast of Southern England—the World Heritage Jurassic Coast and more* (http://www.southampton.ac.uk/~imw/westpubl.htm), provides a comprehensive bibliography of the Jurassic Coast. Updated information on the area is provided in a special issue of the *Proceedings of the Geologists' Association of London*, which appeared in 2019. Details of important sources on both the general geology of the area and individual geologists are given in the references at the end of this book.

The rocks, in almost unbroken succession, have combined to make the exposures of the Devon and Dorset coasts a key training ground for geologists. The first published recognition of their geological value was recorded by the naturalist John Ray (Essex man, pastor/naturalist, and Fellow of the Royal Society, 1627–1705) who in 1673 (p. 115) noted, very much in passing, 'Lyme in Dorsetshire' as a source of 'serpent-stones' gathered by himself and friends. Ever since, many of the founding scientists of the subject have cut their teeth on these sections or been trained on them. British and world museums are full of high-quality material collected from the coast. Many new fossils have been named from fossils found in the region, and, for example, Lyme Regis itself has yielded type specimens of at least 50 species of fossil fish and of 14 fossil reptiles (Dinely and Metcalf 1999; Benton and Spencer 1995).

The rock strata exposed along the Jurassic Coast provide an almost unbroken record of Earth history between 251 and 66 million years ago in relatively un-deformed sediments, representing a remarkable range of past environments. The structure of the coast displays its geological interest superbly. In general, the strata dip gently to the east into the Wessex Basin. The oldest rocks are therefore found in the west, with progressively younger strata outcropping to the east. As a direct result, most parts of the succession are readily accessible in sequential order within the cliffs and foreshore, while the ongoing processes of coastal erosion mean that the exposures are constantly refreshed and new material is brought to light. As it has been said, a walk along the coast is a walk through deep time (see Brunsden [2003] for a general introduction).

The nature of the rocks encountered has depended in part on the latitudinal position of the British Isles. This has changed as the positions of the Earth's plates, continents, and oceans have changed through time (Figure 1.5).

Taking the oldest rock first, the Permian-Triassic boundary, which has always been the subject of some diversity of views (Gallois 2019), is now drawn at the base of the Budleigh Salterton Pebble Beds, making the Aylesbeare Mudstone Group Permian in age. The Aylesbeare Mudstone Group covers the Capitanian and Wuchiapingian stages of the Permian. The Budleigh Salterton Pebble beds are now officially named the Chester Formation, and the Otter Sandstone is the Helsby Sandstone Formation.

The Triassic succession (Figure 1.6) is a virtually continuous exposure of *c.* 1,100 metres of sediments representing most of the Triassic period (*c.* 251–201 million years ago). It consists of continental, terrestrial red-beds and,

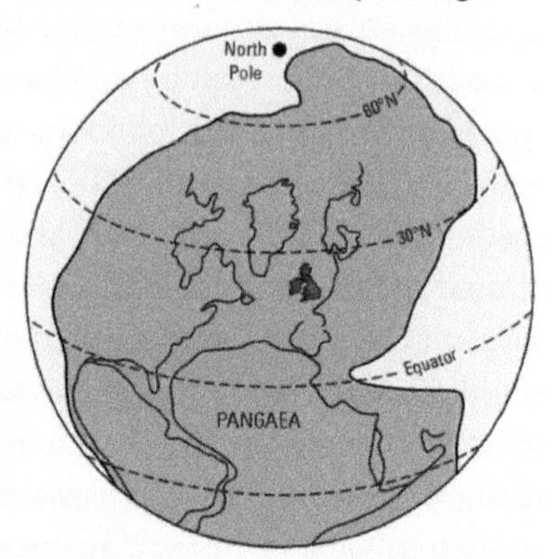

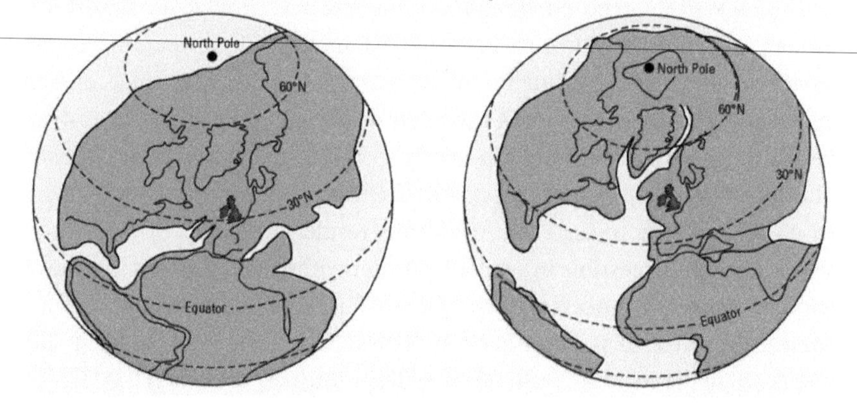

Figure 1.5 The changing latitudinal position of the British Isles (shown in red) through geological time (ASG).

near the top of the sequence, a shallow marine facies. These exposures record evidence of (i) the gradual destruction and denudation of mountains formed in the Variscan orogeny (a mountain-building event at c. 290 million years ago) caused by Late Palaeozoic continental collision between Euramerica (Laurussia) and Gondwana to form the supercontinent of Pangaea at 330–280 million years ago and (ii) the establishment of a widespread marine environment within a Jurassic basin formed during the opening of the Atlantic Ocean. At the time of deposition, this area was located 10–15° N of the Equator. The Mid-Triassic Otter Sandstone Formation (now known as the Helsby Sandstone Formation) at High Peak and Otterton Point in East Devon has yielded species of reptiles, fish, and amphibians (Hart 2014).

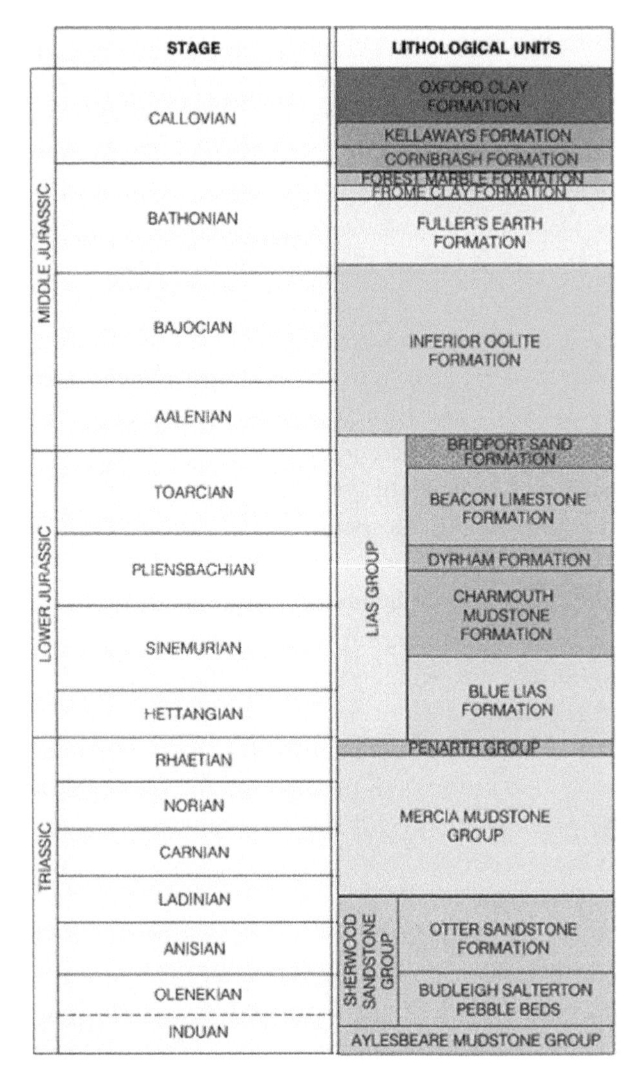

Figure 1.6 Generalised geological succession from the Triassic upwards (modified from © Dorset County Council, 2000, Nomination Document, figure 4). This figure shows the boundaries and nomenclature at the time of the document. See the text for more recent divisions.
With permission of the Jurassic Coast Trust.

It has a lower part formed of aeolian sands, and these pass upwards into a series of river channel deposits. These river channels, which are best seen between the mouth of the River Otter and Ladram Bay, contain a series of classic fluvial facies, such as erosive channel bases, braided stream features, and channel lag deposits. These are magnificently exposed, along approximately 10 kilometres of sea cliffs and intertidal foreshore ledges extending eastwards from Budleigh Salterton to just east of Sidmouth (Coram et al. 2019). Edwards (2008) provides an excellent introduction to the geology and landforms of 'The Red Coast' between Exmouth and Lyme Regis, while Campbell (2006) provides a guide to the undercliffs of the Axmouth to Lyme Regis Nature Reserve.

The rocks along the Jurassic Coast provide one of the finest marine sequences of this age anywhere in the world: every stage of the Jurassic is represented (Figure 1.7). The succession provides excellent evidence of the history of the Earth between *c.* 200 and 146 million years ago, recording six major cycles of sea level change represented by repeated rhythms passing from clay to sandstone and then limestone. These sections have played a key role in the establishment of modern stratigraphy and of biostratigraphic studies. The Liassic vertebrates of Lyme Regis are world famous.

The Lower Jurassic Lias succession of the Jurassic Coast is the best-exposed in Europe (see Lord and Davis [2010] for a discussion of the fossils). It is exposed in an almost continuous cliff section extending from Pinhay Bay to the west of Lyme Regis to Burton Bradstock approximately 15 kilometres to the east. Equally, the Kimmeridge-Portlandian succession along the Dorset coast is possibly the best continuous exposure of rocks of this age in the world. In Kimmeridge, the remarkable Etches Collection shows off a wonderful collection of fossils from that area. The Isle of Portland (Figure 1.8) is the type locality for the Portlandinan (Tithonian Stage), and the cliff sections contain magnificent sections of international importance for stratigraphy, palaeontology, and facies analysis. Portland Stone has been much quarried (and now mined) and has been used all over the world (Hackman 2014; Godden 2016; Butler-Warke and Warke 2012).

The boundary between the Jurassic and Cretaceous probably lies within the lowest beds of the Purbeck Group (Ensom 2007). The Purbeck Group on the Isle of Purbeck (including those rocks that straddle the Jurassic-Cretaceous boundary) contains an exceptional assemblage of vertebrate fossils. Fish remains are locally and exceptionally well preserved. There are also superb remains of Late Jurassic fossil forests exposed on the Isle of Portland (Figure 1.9) and the coast of Purbeck. These forests once grew on the margins of a large hypersaline lagoon, or *sabkha*, which existed 140 million years ago. It is a uniquely complete record of a forest of this age and contains large trees, sometimes *in situ*, with associated algal burrs, which formed around the bases of the trees, and fossilised soils (the so-called *dirt beds*) and pollen. The wood is exceptionally well preserved in silica, displaying microscopic details

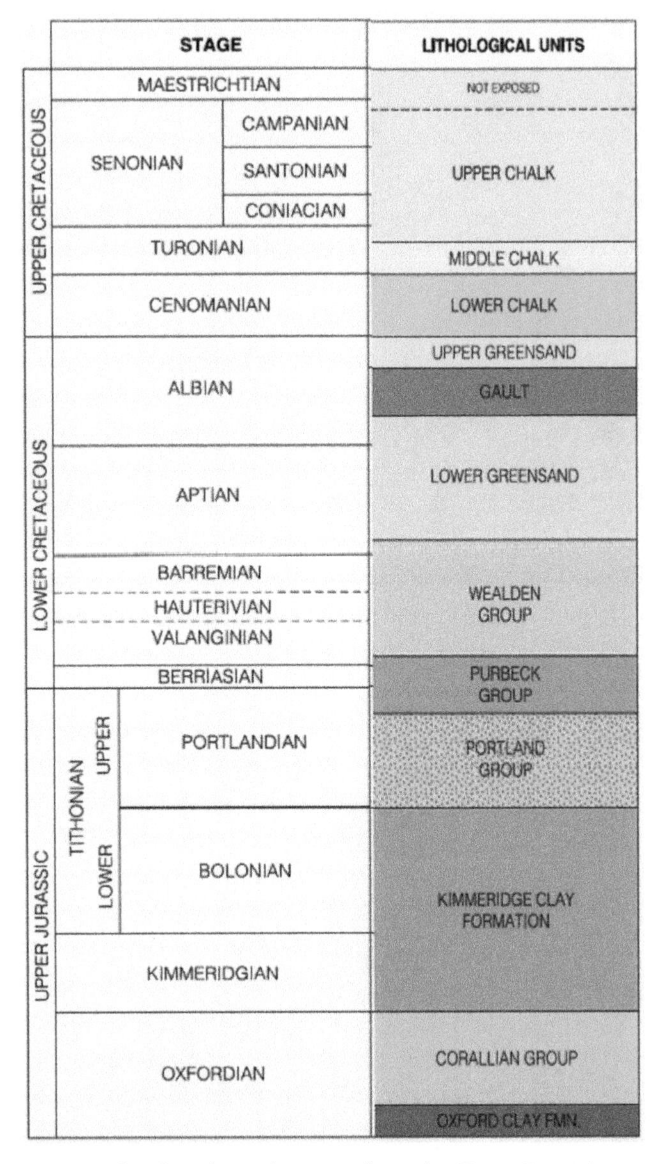

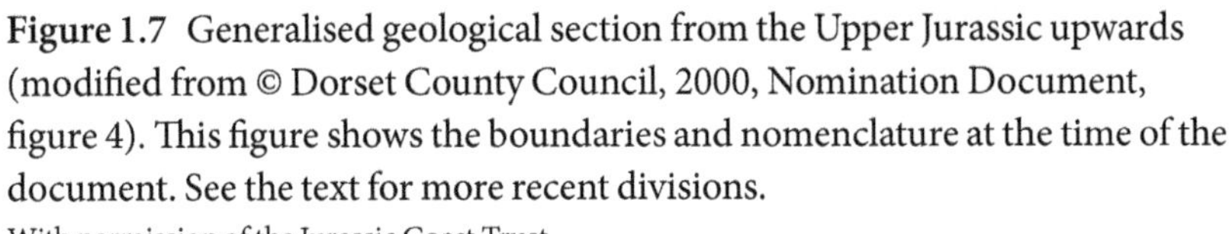

Figure 1.7 Generalised geological section from the Upper Jurassic upwards (modified from © Dorset County Council, 2000, Nomination Document, figure 4). This figure shows the boundaries and nomenclature at the time of the document. See the text for more recent divisions.

With permission of the Jurassic Coast Trust.

Figure 1.8 The West Weares of the Isle of Portland showing the Portland Beds (ASG).

Figure 1.9 Fossil tree trunk from the basal Lulworth Formation of the Purbeck Group at the Portland Heights. Amy Beasley (nee Goudie), for scale (ASG).

of the structure, including preserved growth rings which allow a detailed assessment of the climate of the time. The Purbeck Group Fossil Forest is one of the most complete examples of a fossil forest of any age and, unlike many others, preserves *in situ* trees with soils.

The fossil forest at Lulworth and on Portland is a very primitive-looking landscape of thrombolites (Figure 1.10). These thrombolites (Bosence and Gallois 2021), formed by bacteria, look like rock doughnuts bored through with deep craters where trees were once rooted. They were formed when a forest of trees akin to cypresses was growing in a swamp on the edge of a warm lagoon or saline lake in a Mediterranean-type, semi-arid environment. The forest was later drowned by the sea, and, over many millions of years, its remains were preserved within layers of limestone that were deposited on the seabed and then compressed. There are some logs and trunks which were fossilised, but, for the most part, the trees rotted away, leaving just the hollow mounds of the algae (stromatolites) that had formed around their trunks.

The overlying Wealden Group (Penn et al. 2020), along with that on the Isle of Wight, is one of the most complete sequences of this age available at a single site in north-west Europe. Durlston Bay near Swanage yields the richest assemblage of Lower Cretaceous insects, fish, early mammals, and

Figure 1.10 A thrombolite from King Barrow (Kingsbarrow) Quarry, at the north end of Portland. It is about 1 metre across. The central hole once contained a tree stump, rooted in the Great Dirt Bed palaeosol below (ASG).

reptiles known anywhere on Earth. These were made famous through the researches of Samuel Beckles and Peter Brodie. The succession within the Jurassic Coast includes rocks of all stages of the Cretaceous period, with the exception of the uppermost one. The Cretaceous Chalk is well exposed (Mortimore 2019), particularly in the cliffs in the east of the area between Weymouth and Lulworth, and at Ballard Down. The old division of the Chalk into lower, middle, and upper has been replaced by the following formations in Dorset. From oldest to youngest these are Basement Bed, Zig Zag Chalk Formation, Holywell Nodular Chalk Formation, New Pit Chalk Formation, Lewes Nodular Chalk Formation, Seaford Chalk Formation, Newhaven Chalk Formation, Culver Chalk Formation, Portsdown Chalk Formation, and Studland Chalk Formation.

Also important are the Upper (Gallois and Owen 2019) and Lower Greensands (Ruffell and Batten 1994). The former plays an important role in landslide development, has in places been karstified, and has generated much chert debris that mantles slopes (Gallois 2005).

Locally in the area there are deposits of Paleogene or Neogene age, including Eocene gravels, sarsen stones, solution hollow (doline) infills, and the like. There are also some major sites of Pleistocene age, including the famous raised beach at Portland Bill, although this area lay beyond the southern limits of the Quaternary ice sheets and thus was subjected to periglacial rather than glacial conditions. Dry valleys (Figure 1.11) in the Chalk (notably Scratchy Bottom to the west of Lulworth; Figure 1.12), which no longer have flowing water, may have been active during these cold conditions. Sea level rise of over 120 metres followed the end of the last Ice Age (from around 20,000 years ago), occurring from the late Pleistocene and through the most recent period of geological time, the Holocene. This brought the coastline to its present position and drowned an extensive offshore landscape.

The Jurassic Coast was inscribed by the World Heritage organisation of UNESCO, not only for its geological sequence and fossils, but also for the quality, variety, and significance of the landforms and ongoing geomorphological processes (Goudie and Migoń 2020; Allison 2020). This derives mainly from the interaction of oceanic and atmospheric controls on the continuous variation of the lithology of the eastwardly dipping strata. Some of the landforms are amongst the most frequently quoted examples in textbooks.

The coast is iconic for the study of landslides, with a huge variety of forms and dramatic changes. Famous landslides include the Hooken slide, near Bere, of 1790; the Southwell topple of 1734 on Portland (Dykes and

Figure 1.11 The Cretaceous chalk cliffs of the Dorset coast west of Durdle Door, serrated by dry valleys (ASG).

Figure 1.12 Scratchy Bottom, near Lulworth. A great dry valley excavated into the Chalk under periglacial conditions (ASG).

Bromhead 2021); the East Weares slide on Portland in 1792; and the Bindon slide of 1839 in East Devon. The last of these was made famous by the work of Dean William Conybeare and Dean William Buckland (see Lincoln in https://www.lymeregismuseum.co.uk/lrm/wp-content/uploads/2017/02/10_views_fiinal_version_-21_-June_-2021.pdf). The Axmouth to Lyme Regis undercliff is the only landslide National Nature Reserve in the United Kingdom (Pitts 1979, 1983). This piece of the coast has been the subject of considerable research in the past 50 years, with many innovative advances in image interpretation and monitoring. Landslips have occurred near Sidmouth (Gallois 2011) and Salcombe Regis (Gallois 2007). The landslip complexes between Lyme and Charmouth include Black Ven and the Spittles (Gallois 2014), while to the east of Charmouth lies Stonebarrow (May 2003a). There is a large landslide complex between Ringstead and White Nothe, and large mudflows have occurred at Osmington. Mass movements have also occurred elsewhere: for example, Burton Bradstock, in July 2012; St Oswald's Bay near Durdle Door, in May 2013; Bowleaze Cove near Weymouth, in April 2016; Worbarrow Bay, in May 2017 (Figure 1.13); West Bay, in March 2019; Eype, in November 2020; Hive Beach, in December 2020; and between Seatown and Eype, in April 2021.

The Isle of Portland is one of the most instructive places in the world to demonstrate the influence of geological structures on cliff processes and forms (Goudie 2020; Figure 1.14). In 1836 (p. 226), W. H. Fitton, an eminent gentleman geologist wrote: 'Few places, it is probable in the world, exhibit with such distinctness and in so small a place, phenomena of more extraordinary interest, or of greater importance to theory.' Indeed, the whole Jurassic Coast has dramatic and very beautiful cliffs of national scenic importance. More than 80% of the length of the East Devon and Dorset coast is composed of cliffs developed in sedimentary rocks. Stacks and arches at Ladram Bay, Stair Hole, Durdle Door (Figure 1.15), and Old Harry Rocks, and arcuate bays like Lulworth Cove (Nowell 1998), Worbarrow Bay (Nowell 2000), and Chapman's Pool, all feature in most guidebooks and school texts. The Purbeck coast perfectly illustrates the principles of concordant and discordant coastline evolution models.

At the foot of many cliffs are striking examples of shore platforms, as at Charmouth/Lyme, Portland Bill, Newton's Cove (Weymouth), and Kimmeridge.

The area is one where the power of humans to change coastal processes and erosion rates can be demonstrated. This is the case, for example, at Sidmouth

Figure 1.13 A great landslide at Worbarrow Bay, in May 2017.
Source: © Google Earth (2018).

Figure 1.14 The West Weare cliffs of Portland taken from the Cove House Inn (Alice Goudie); a superb place to see the influence of rock characteristics and structures on cliff forms.

Figure 1.15 Durdle Door, one of the iconic landforms of the Jurassic Coast (ASG).

(Gallois 2011) and at West Bay, where harbour works have upset the natural pattern of erosion and deposition.

The beaches vary in composition, from the red quartzite pebbles at Budleigh Salterton, to the cherts and flints of West Dorset or the sands of Weymouth and Swanage. The sediment budgets for the coast of West Dorset have been analysed by Bray (1996). Chesil Beach (Figure 1.16), studied by Sir John Coode, extends for 28 kilometres from Bridport Harbour to Portland,

Figure 1.16 Chesil Beach and the Fleet in the evening sun (ASG).

becoming steeper and higher from west to east (May 2003b). It is separated from the mainland by the Fleet, which contains important sediments recording the sea level and climatic changes of the past 7,000 years. This is also Europe's largest coastal lagoon, while the beach is famous worldwide for the volume, type, and size-grading of its pebbles. Papers concerning the origin and processes of the beach have been written for more than 200 years and are still controversial.

The off-shore submarine geomorphology is an area of burgeoning interest and contains some remarkable features, including circular structures, sand accumulations, and ancient river-courses (May 2019).

Box 1.2 The Role of Women at This Time

One matter of interest that has arisen from writing this book has been the role of women. Most reviews and contemporary comments on the contributions of women to the Earth Sciences suggest that, following the mores of society in the nineteenth century, they were often born into influential, scientific, and well-off families where the man of the house had independent means and training, employment, or amateur interests, usually in botany, astronomy, or palaeontology. This allowed, or ensured, that many women worked voluntarily with no recognised status or salary. They were often described as assistants, preparators, illustrators, curators, or secretaries. Although many women discovered specimens or made fundamental insights and original scientific observations, they generally received little scientific credit. Men who worked with women—often their wives, sisters, or daughters—were described as 'extraordinarily enlightened'.

Many women who were mothers, governesses, or teachers concentrated on education for young people, other women, and for the general public. Some wrote academic papers, but it was more common to write in popular texts, such as *Every Girl's Magazine.* They also wrote biographies of the men with whom they were associated and completed their works posthumously. At the same time, their own work was often anonymous, with their substantive work appearing under a man's name. The discoveries of Mary Anning, such as the plesiosaur, written by William Conybeare, is the most startling example.

The nineteenth century is noted for the development of secondary and university education for ladies. The London Collegiate School, Cheltenham's Ladies College, Queens College, and the Girl's Public Day School Trust (1871–1878) are important milestones. Girton and Newnham in Cambridge, Somerville and Lady Margaret Hall in Oxford, and Birkbeck University and King's College in London all played a distinguished role. In geology two academics stand out: Catherine Raisin at Bedford and Gertrude Elles at Newnham, but neither is associated with the Jurassic Coast.

The scientific community did not support women. Learned societies, such as the Royal Society and the Geological Society, did not admit women as Fellows until 1917, by law. Fortunately, local or regional groups, such as the natural history, scientific, and literature societies of the country, were more liberal, particularly if they were based on outdoor learning and collecting, which also might include collecting and observation by children and healthy pursuits. This was because botany was seen as suitable for children and ladies.

The value of natural history and outdoor pursuits is perhaps why the Geologists' Association, based on amateur scientists, had equal rights from its foundation in 1858. The Geographical Association, developed mainly by teachers, had a similar history at a time when Fellowship of the Royal Geographical Society, under the leadership of Lord Curzon, was strongly against female membership.

Even then, however, the situation was patronising although well-intentioned. Men were widely regarded as superior and better able to understand science, be objective, and to reason. As Thomas Jones, FRS (1819–1911), President of the Geologists' Association, said in his opening address to the 1880–1881 session,

> Women as well as men, can be Geologists as far as their strength for travel and opportunities among domestic affairs will allow. Doubtless for many ladies it is hard to tramp about on Geological Excursions over rough roads, hillsides, hedges, ditches and seaside rocks and shingle. . . .
>
> Let special Excursion lines be planned so that we may have the pleasure and advantage of female society. . . .

They have often graced the outings with their presence. . . .

> In the meetings of the Association the female element is an adornment and a social pleasure . . . how greatly would the scientific meetings and outings lose if the ladies were absent! . . .
>
> Who requires to be reminded of the intellectual powers and persistent energy of the well-educated woman?

Although many of these women are still described as 'unusual for their time' and were certainly few in number, the picture for the Jurassic Coast is unusual, perhaps because Mary Anning stands out as an independent figure in the history of science. Her discoveries were reported and scientifically described by men. She resented this, but later they acknowledged her abilities and contribution. Some supported her financially and, after her death, subscribed to a memorial in her name, the stained-glass window in Lyme Regis Parish Church. Sir Henry De la Beche, founder of the Geological Survey of Great Britain, wrote a fulsome obituary. In Lyme Regis, the Philpot sisters were also great collectors, and the town's museum bears their name. There are a few important other figures that we have reported here, including Mary Buckland, Marie Stopes, Eleanor Coade, and Muriel Arber.

Box 1.3 Local Museum Resources

Readers of this book should visit some of the excellent museums along the coast that have fossil displays. These include Sidmouth Museum, Seaton Jurassic, Lyme Regis Museum, the Charmouth Heritage Coast Centre, Bridport Museum, Portland Museum, The Etches Collection Museum of Jurassic Marine Life in Kimmeridge, and the Swanage Museum and Heritage Centre. Also containing important material from the coast are the Dorset County Museum in Dorchester and the Royal Albert Memorial Museum in Exeter. The Charmouth centre was awarded the R. H. Worth award of the Geological Society of London, in 2019, for its outreach, public engagement, and role in education.

Box 1.4 What is Geomorphology?

The authors are both geomorphologists. The word 'geomorphology', which means literally to write about (Greek *logos*) the shape or form (*morphe*) of the earth (*ge*), first appeared in 1858, in the German literature. The term was referred to by the French geographer Emmanuel de Margerie, as *la géomorphologie* in 1886. It first appeared in English in 1888, and it was used at the International Geological Congress in 1891. As the British Society for Geomorphology has stated on its website (https://www.geomorphology.org.uk/what-geomorphology-0),

> Geomorphology is the study of landforms, their processes, form and sediments at the surface of the Earth (and sometimes on other planets). Study includes looking at landscapes to work out how the earth surface processes. . . . Landforms are produced by erosion or deposition, as rock and sediment is worn away by these earth-surface processes and transported and deposited to different localities. . . .
>
> Earth-surface processes are forming landforms today, changing the landscape, albeit often very slowly. Most geomorphic processes operate at a slow rate, but sometimes a large event, such as a landslide or flood, occurs causing rapid change to the environment, and sometimes threatening humans. . . .

It is also important to appreciate the role of tectonics in creating landforms.

Geomorphologists are also 'landscape-detectives', working out the history of a landscape. Most environments, such as Britain and Ireland, have in the past been glaciated on numerous occasions, tens and hundreds of thousands of years ago. These glaciations have left their mark on the landscape, such as the steep-sided valleys in the Lake District and the drumlin fields of central Ireland. Geomorphologists can piece together the history of such places by studying the remaining landforms and the sediments . . .'. However, we should mention that the World Heritage Coast itself was not glaciated in the Ice Age.

2
The Earliest Investigators

A prime reason for the World Heritage inscription of the Jurassic Coast is the role its features have played in the history of science. The site has been studied for more than 300 years, acquiring particular fame in the early part of the nineteenth century, a critical time for the development of many of the fundamental ideas of the Earth (see Box 2.1).

The story we tell is how the natural world of landforms, processes, convulsions of the Earth, the evolution of life, and the nature of rocks was discovered in one of the birthplaces of Earth Science—the coasts of East Devon and Dorset. It is about the lives of both the scientific and amateur men and women who observed, collected, recorded, described, theorised, and published their discoveries.

These extraordinary people were acting within a framework of belief that was largely determined by the story of Creation as given in the Christian Book of Genesis. For the early workers on the Jurassic Coast, the dominant belief was that the world was created in a short period of time and that the features we saw now had remained largely unchanged. It was the handiwork of God. They recognised that 'unnatural forces' such as 'the universal deluge', Noah's Flood, earthquakes, or other large events had occurred since the initial creation, and many explanations were given on what the dominant controls and changes were. However, underlying all of these ideas was the belief that there had been a single act of creation: it was a principle and a religious truth.

This provided the early observers with a fundamental problem. They observed evidence of earlier life forms, animal and plant remains in places where they should not be (e.g. sea shells on the top of mountains), clear signs of current erosion and deposition, and obvious landform change as the forces of nature attacked rocks with varying frequency and magnitude. Many of the leading lights who discussed these problems worked on the Jurassic Coast, and, as we follow their progress, we move through the development of theories of science and the theological understanding of the meaning of Creation. In the end, serious observation of incontrovertible facts overcomes dogma.

Geological Pioneers of the Jurassic Coast. Andrew S. Goudie and Denys Brunsden, Oxford University Press.
 DOI: 10.1093/oso/9780197638088.003.0002

Box 2.1 Some Key Terms

The terms 'geology' and 'scientist' are relatively recent ones. The term 'geology' was first used technically in publications by two Genevan naturalists, Jean-André Deluc and Horace-Bénédict de Saussure, though 'geology' was not well received as a term until it was taken up in the very influential compendium, the *Encyclopédie*, published beginning in 1751 by Denis Diderot. The term 'scientist' was only invented in 1834, by the Cambridge scholar William Whewell. The term 'palaeontology' (paléontologie) was first used in 1822 by Henri Marie Ducrotay de Blainville, one of Cuvier's former students, to refer to the study of ancient living organisms through fossils. In 1829, the French naturalist Alexandre Brongniart published a survey on the different terrains that constitute the crust of the Earth and referred to the terrains of the Jura Mountains as *terrains jurassiques*, thus coining and publishing the term for the first time.

It is not possible to discuss here the deep foundations of the science attributed to such great minds as Aristotle (384–322 BC), Leonardo da Vinci (1452–1519), Girolamo Fracastoro (1476/8–1517), George Bauer (1494–1555), Bernard Palissy (1510?–1590), or Nicholas Steno (1638–1686). Our story really begins with an almost unknown figure, a Weymouth customs officer called William Hobbs. In 1715, he wrote: 'ye manner how, and when, the Shells, and other marine productions, came to be immassed and mingled in the Rocks and Mountains'.

In the early eighteenth century he was a significant contributor to our knowledge of the geology of the Jurassic Coast and held views on the origin of sediments and fossils that were at variance with the orthodoxy of the time. He rejected the explanation of Noah's Flood and believed that fossils were the remains of real animals that had been formed in the environments that formed rocks on the sea floor. He was 'way ahead of his time. He remains an almost unknown figure who worked at the end of the seventeenth century.

Our second great mind is Robert Hooke. In 1668, at the dawn of the Enlightenment, he stood on the construction site of the Royal Exchange, inspecting blocks of Portland stone. He observed that the blocks were pitted and veined with strange geometries, concluding that these shapes must be the traces of creatures that had once inhabited the Earth yet had subsequently vanished. This was perhaps the first realisation of extinction. Widely travelled, he collected fossils on the coast and was perhaps the first to conclude, from his observations of shell-like fossils, that shells could also be turned into fossils when they were 'fill'd with some kind of Mud or Clay, or petrifying Water, or some other substance'. He was also the first to study them with a microscope.

Our third important figure is John Woodward, a professor of physics at Gresham College, who collected fossils on the south coast and assiduously recorded the 'Entrails of the Earth'. He revealed that terrestrial matter could be distinguished into strata and that these contained shells of once living animals. Because of the religious tempo of the day, he was forced to use a catastrophic explanation based on the deluge, which had reordered the features of Earth and redeposited them into layers of different density as the flood waters receded.

At this time we also have James Parkinson, a famous surgeon, but also one of the most enlightened geologists and palaeontologists of the eighteenth century. He wrote an illustrated book, *Organic Remains of a Former World* (1804–1811) that included many fossils from the Jurassic Coast. Although he believed that fossils 'displayed the munificence of the Lord', he struggled with stories of the Creation and deluge, glimpses of which is revealed in his work.

The struggle to reconcile the account of the Creation with these new observations and collections now becomes a central preoccupation of geologists and churchmen for the rest of the century. We enter what has been called 'The Golden Age of Geology'. It is a story in which our knowledge moves from theological ideas of creation to an understanding based on scientific, observable fact and the development of powerful scientific principles, theory, and pragmatism. The grand debates between protagonists of the universal ocean (Abraham Werner), catastrophism (William Buckland), uniformitarianism (James Hutton), and the importance of time (Charles Lyell) were in part played out by their followers on the Jurassic stage.

The story also took place against a turbulent social backdrop. The intellectual background was favourable to our science. This was the eve of the French Revolution when Rousseau (1762) wrote 'Man is born free but is everywhere in chains.' It also saw the abolition of the Slave Trade Act (1807), the emancipation of Catholics (the Roman Catholic Relief Act of 1829), and the Nature School of Education promoted by Johann Heinrich Pestalozzi (1746–1827). The intellectual atmosphere was tending towards trying to create liberty of mind, freed from the religious dogma derived from the Middle Ages. The Industrial Revolution was asking for the development of reliable knowledge of chemistry, physics, mathematics, and engineering, such as mineral mining (Abraham Werner, 1749–1817) and canal engineering that required detailed mapping, correlation of strata, and structural models (William 'Strata' Smith, 1769–1839).

Established religious beliefs were under attack from the Dissenters, Congregationalism (e.g. Mary Anning's family), and Methodism so that the established church was questioned. Public education was weak, and universities were mainly for the education of young men. Politicians studied classics, if at all. Young country gentlemen entered the church, army, or government.

This was, however, also the time of 'the Study of Nature', when natural history was seen to be a profitable activity, healthy and related to the collecting mania of the day and the creation of the great museums that were to characterise the Victorian era. There was a development of interest in science in general, and, in the 1660s, the Royal Society was established in London. Some of its founding fathers, such as Robert Hooke, John Ray, and John Woodward, had an interest in matters geological. A medical doctor who worked briefly on the geology of the Jurassic Coast was the Swiss Jean-François Berger (1779–1833). John Middleton (1751–1833), essentially a surveyor and agriculturalist, helped to establish the stratigraphic sequence in Surrey and Dorset in 1812–1813.

William Hobbs

In the early eighteenth century, William Hobbs was a significant contributor to our knowledge of the geology of the Jurassic Coast. He held views on the origin of sediments and fossils that were at variance with the orthodoxy of the time. He did not accept that the source of sediments was a result of the biblical deluge (Noah's Flood). Instead, he believed that the fossils found in the local rocks were the remains of once real animals which had not been emplaced by the flood (Figure 2.1) but had been embedded in rocks which had formed on the sea floor. He also found that these rocks had later been raised up by 'pulsations from the centre of the Earth'. His views seem to have been stimulated by his study of the present formation of the Chesil Bank and of the fossils exposed in the cliffs of the Isle of Portland. These ideas mark him out as one of the most important and prescient thinkers in the science.

As Roy Porter showed in his biography of Hobbs and in his edition of Hobbs's *The Earth Generated and Anatomized* (*c.* 1715), Hobbs was a

Figure 2.1 Francis Danby's depiction of Noah's flood, c. 1840. Until the mid-nineteenth century, many geologists believed that the Flood explained many phenomena, including the deposition of fossils and the excavation of valleys. Hobbs was an early exception.

Source: https://commons.wikimedia.org/wiki/File:Francis_Danby_-_The_Deluge_-_Google_Art_Project.jpg.

shadowy figure. It is not clear when he was born, though it was probably before c. 1670. There is also 'no unequivocal clue as to his upbringing, education or vocation training' (p. 6). He seems to have been sacked in 1705 by the Board of Excise for dishonesty. He also seems 'to have lived a life of almost continuous isolation from the learned and philosophical communities of the day' (p. 9). It is not clear when he died.

Robert Hooke

Robert Hooke (1635–1703), born on the Isle of Wight, was an English natural philosopher, architect, and polymath. He studied at Wadham College, Oxford. At one time he was simultaneously the curator of experiments of the Royal Society, a member of its Council, Gresham Professor of Geometry, and Surveyor to the City of London after the Great Fire. He was also an important architect. He was said to be irascible, at least in later life, proud, and prone to take umbrage with intellectual competitors. He spent his life largely on the Isle of Wight, in Oxford, and in London.

Hooke's major geological contribution, *Lectures and Discourses of Earthquakes and Subterraneous Eruptions* (1705), represents the contents of a series of lectures that he probably presented at the Royal Society in 1667 and 1688. He had come into the possession of a large number of fossils from southern England, including ammonites (Figures 2.2. and 2.3) and echinoids. On page 327 of *Lectures and Discourses*, he referred to the 'Snake or Snail

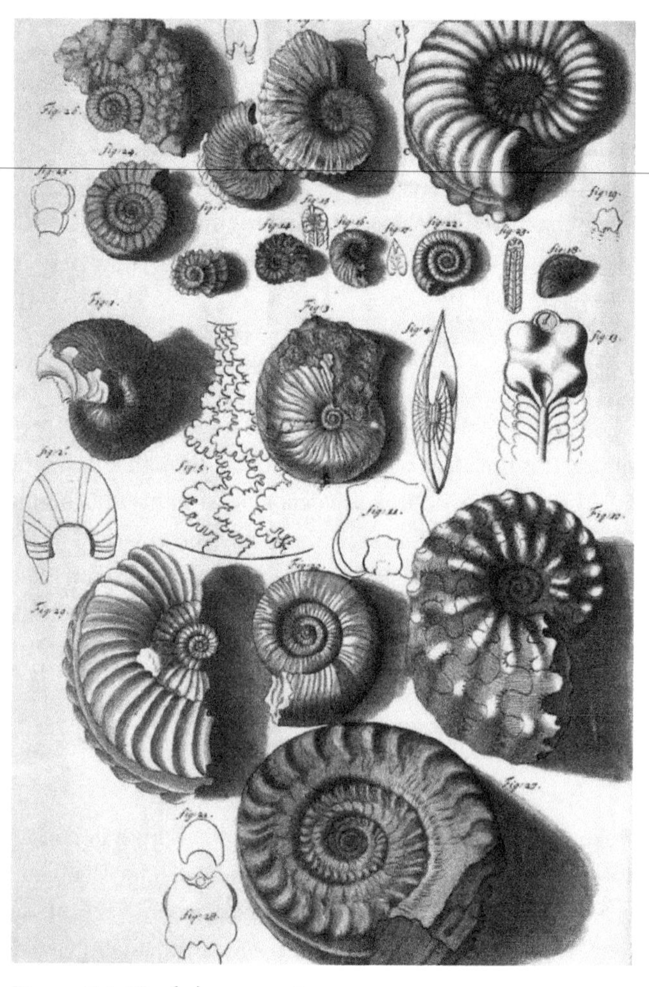

Figure 2.2 Hooke's ammonites.
Source: Hooke (1705).

Figure 2.3 Robert Hooke.
Source: https://makingscience.royalsociety.org/s/rs/people/fst00009590.

Stones' of Portland—what we now term ammonites—and noted that there were great varieties of them, some of which 'were of a prodigious bigness'.

Having been exposed to this material, Hooke was faced with an intellectual problem. How could it be that these shells, which were evidently the remains of once living marine organisms, were frequently found buried considerable distances above sea level and sometimes many miles from the sea?

He did not think, as was the current feeling, that they were mere 'sports of nature'. He believed that earthquakes had caused changes in level and that long continued denudation had dumped debris into the sea in which the fossils occurred. Hooke examined fossils with his microscope, thus becoming the first recorded person to do so. Through his observations, he noticed striking similarities between petrified and living wood, and fossil shells and living mollusc shells. Comparing petrified wood to a piece of rotten oak wood, he realised that wood could be turned to stone when water deposited minerals throughout it. Similarly, his observations on shell-like fossils led him to conclude that shells could also be turned into fossils when they were immersed in clay, sand, or 'petrifying water'. Hooke was not minded to follow the biblical story of the Creation or the biblical chronology, reasoning that the Bible was written by humans and that many things had happened before humans had been created.

In this recognition of the origin of marine fossils, Hooke was considerably ahead of his time.

John Woodward

John Woodward (1665–1728) (Figure 2.4) was one of the earliest people to make sense of fossils. He may be regarded as the foremost British geologist of the era before James Hutton (1726–1797), William Smith (1769–1839), and Charles Lyell (1797–1875). It is thought that he was born in Derbyshire, at some point in the mid to late 1660s. At the age of 16, he went to London, where he was initially apprenticed to a linen draper. Later, he studied medicine with Dr Peter Barwick, physician to King Charles II. In 1692, he was appointed Gresham Professor of Physic; in 1693, he was elected a Fellow of the Royal Society; and in 1702, he became a Fellow of the Royal College of Physicians. He appears not to have been a very good practitioner. His claim that all illnesses could be treated by vomiting earned him the sobriquet of 'Don Bilioso'.

Woodward was not a highly popular individual. Contemporary evidence indicates that his character was not such as would have endeared him to his contemporaries. They were sickened by him. He was ill-mannered, quarrelsome, easily offended, and had a great conceit of himself. Davies (1969, p. 75) reported that 'He was renowned for his eccentricities, pomposity, irritability,

Figure 2.4 John Woodward, drawn in 1774 by W. Humphrey.

Source: https://en.wikipedia.org/wiki/John_Woodward_(naturalist)#/media/File:John_Woodward.jpg.

and bad manners, and his personal vanity was such that he had mirrors fixed in all his rooms so that he might lose no opportunity of gazing at himself.' The antiquarian William Stukeley described him as an 'egregious coxcomb' (Haycock 2002, p. 81). He was expelled from the Council of the Royal Society in 1710, for conduct unbecoming a gentleman. He never married and has been described as 'notoriously homosexual'.

While still a student, he became interested in botany and natural history, and, during visits to Sherborne in Gloucestershire, where he may have had family connections, his attention was attracted by the fossils found there in the Jurassic beds. He began to form his great collection. He published *An Essay toward a Natural History of the Earth and Terrestrial Bodies, especially Minerals, &c.* in 1695. In 1728, a catalogue of the British fossils that Woodward had collected was published posthumously (details are given in Price 1989). Specimen A1 in the list, consisting of portions of the shank of a dinosaur limb bone, is still preserved in the Woodwardian collection at Cambridge. It is perhaps the earliest discovered dinosaur bone still identifiable (Delair and Sarjeant 1975, p. 8).

Subsequent editions followed, and *An Essay* was translated into French, Italian, German, and Latin. This was followed in 1696 by *Brief Instructions for making Observations in all Parts of the World*. He later wrote *An Attempt towards a Natural History of the Fossils of England* (2 vols., 1728 and 1729) (Figure 2.5). In these works he showed that the stony surface of the Earth was divided into strata and that the enclosed fossils were originally generated in the sea. He advocated and emphasised that fossil remains were organic in origin, a view that was not universally accepted at the time he wrote. As Roy Porter wrote in 1979,

> Woodward rejected the commonly held view that these objects were not petrified remains at all, but were rather *lapides sui generis*, authentic stones which, like crystals, had shot into lifelike shapes, on account of the sportive powers of nature. To conceive of nature playing tricks like that on observers, believed Woodward, was unworthy of God and would put an end to all science. . . . But if these extraneous fossils were organic remains, how had they become embedded in solid rock? . . . Woodward's solution to these thorny problems of strata and fossils was the Biblical Deluge.

He gave credence to the Book of Genesis and to the great Flood as causes of change.

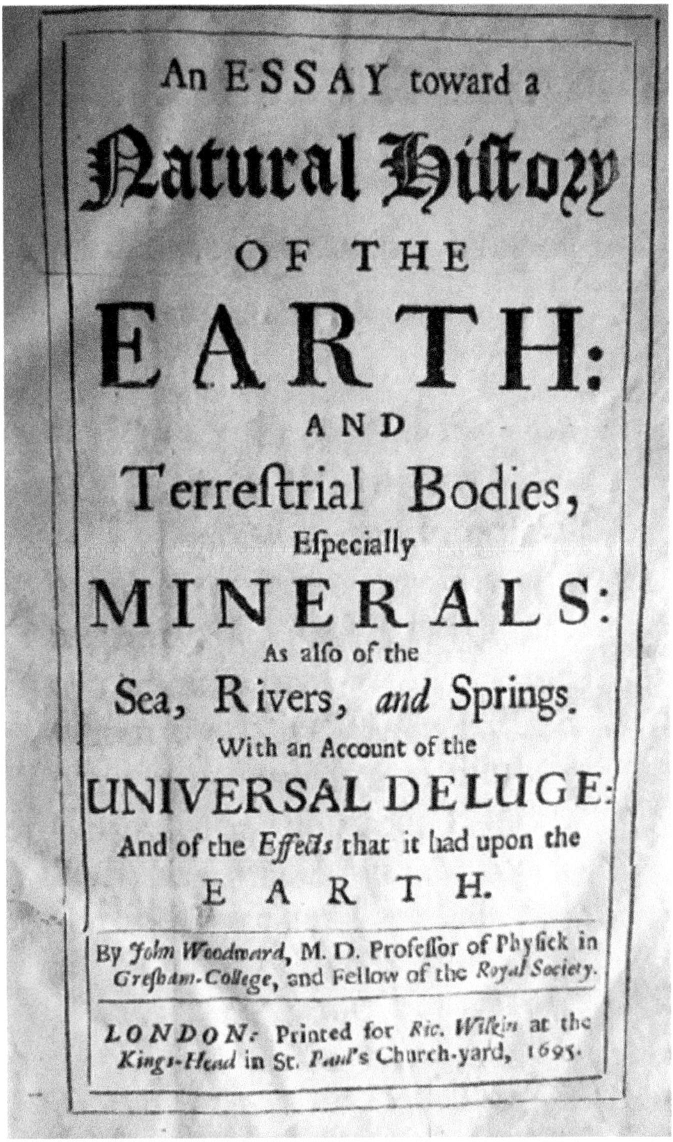

An ESSAY toward a
Natural History
OF THE
EARTH:
AND
Terreſtrial Bodies,
Eſpecially
MINERALS:
As alſo of the
Sea, Rivers, *and* Springs.
With an Account of the
UNIVERSAL DELUGE:
And of the *Effects* that it had upon the
E A R T H.

By *John Woodward*, M. D. Profeſſor of Phyſick in *Greſham-College*, and Fellow of the *Royal Society*.

LONDON: Printed for *Ric. Wilkin* at the *Kings-Head* in St. *Paul's* Church-yard, 1695.

Figure 2.5 The frontispiece of Woodward's essay of 1695 (ASG).

He was probably strongly influenced by Robert Hooke, his fellow Gresham College professor. Hooke had strongly argued in *Micrographia* (1665) that fossils were remains of organisms. Another important influence was the Sicilian artist and naturalist Agostino Scilla, who wrote, in 1670, a book which argued for the organic nature of fossils (*La Vana Speculazione Disingannata dal Senso* or 'Vain Speculation Undeceived by Sense').

Woodward died on 25 April 1728, and his grave is close to those of Isaac Newton, Charles Darwin, and Charles Lyell in Westminster Abbey. There is an elaborate white marble sculpture in the nave which carries this text (translated from the Latin): 'Sacred to the memory of John Woodward, a doctor the most celebrated, and a philosopher the most exalted, whose ability and learning his writings diffused over the face of nearly the whole Globe, but whose liberality and affection for his country and the University of Cambridge, enriched by his munificence and embellished by his wealth, declares to perpetuity.' In his will, Woodward directed that his personal estate and effects were to be sold and land of the yearly value of £150 purchased and conveyed to the University of Cambridge. A lecturer was to be chosen and paid £100 a year to read at least four lectures a year. This created the Woodwardian Professorship of Geology. He also bequeathed his collection of English fossils to the University, to be under the care of the geology lecturer, and so formed the nucleus of the Woodwardian Museum (the specimens have since been removed to the new Sedgwick Museum of Earth Sciences). His collection of fossils and minerals is unusual for the time in not being a miscellaneous assemblage of the rare and curious but a representative and truly 'scientific collection'. In contrast to the vast majority of collectors at that time, Woodward did not concentrate on 'curiosities', but on typical samples of rocks, minerals, fossils, and the like. Amassed between 1688 and 1724, his collection amounted to around 9,400 specimens. He took great care to record exact localities and details of occurrence for his specimens, which was not a universal practice at the time (Price 1989). As Torrens (2004) has shown, Woodward had visited the Jurassic Coast, and the catalogue shows that a large number of Dorset localities are represented, including Bridport, Lulworth, Lyme Regis, Melbury Abbas, Poole, Portland, Purbeck, Radipole, Shaftesbury, and Weymouth. One of his large Portland ammonites was given to him by Christopher Wren, who used the incomparable Portland Stone for the rebuilding of St Paul's Cathedral after the Great Fire of London.

Davies (1969, p. 83) concluded,

> Woodward undoubtedly deserves more credit than he has been accorded by historians of science. . . . He should be remembered, rather, as the first geologist to work his way systematically through the English countryside, seeking out exposures, collecting specimens, and building up from

his notebooks a picture of the geology of England which must have been unique in its day. . . . He was well aware that some strata contain characteristic fossils, and had he been able to shed his Mosaic shackles, he might have perceived some of the implications of his discovery. Woodward might in fact have become the founder of modern stratigraphy more than half a century before the birth of William Smith, but the grip of Moses was too strong; instead of becoming the father of English geology, Woodward became the first of a long line of English field-geologists who made valiant attempts to harmonise their observations of Nature with the Mosaic record.

James Parkinson

James Parkinson (1755–1824) was born in Shoreditch and buried there in St. Leonard's Church. He was a radical political thinker, an apothecary, and a surgeon who also developed interests in palaeontology but was best known for the disease, 'shaking palsy', which bears his name (Parkinson's disease). He also made contributions to the study of gout (a scourge of many bibulous geologists), the ruptured appendix, and the management of 'mad houses'.

Parkinson trained at the London Hospital. He was the son of a surgeon, and, in Hoxton, east London, became a general practitioner, later in partnership with his son. At the turn of the nineteenth century he became fascinated by natural things, especially palaeontology and general geology. He started to amass a collection when he was only about 17 years old. On frequent excursions he collected fossils, both plant and animal. He was also one of the 13 founders of the Geological Society, and his exquisitely illustrated *Organic Remains of a Former World* (published in three volumes between 1804 and 1811, Figure 2.6) placed the study of fossils on the scientific map of Britain before the subject even had a name. In essence, *Organic Remains* was the first introductory palaeontological text for public consumption. Several fossils were named after him, including an ammonite, *Parkinsonia parkinsoni* (Figure 2.7). In his collection, Parkinson had a magnificent ammonite from Charmouth, together with petrified wood (p. 419 in volume one of his book) and in volume three he gives a number of examples of fossils from 'Dorsetshire'.

Parkinson was one of the most enlightened geologists working in Britain at the start of the nineteenth century. As Cherry Lewis remarked (2017b),

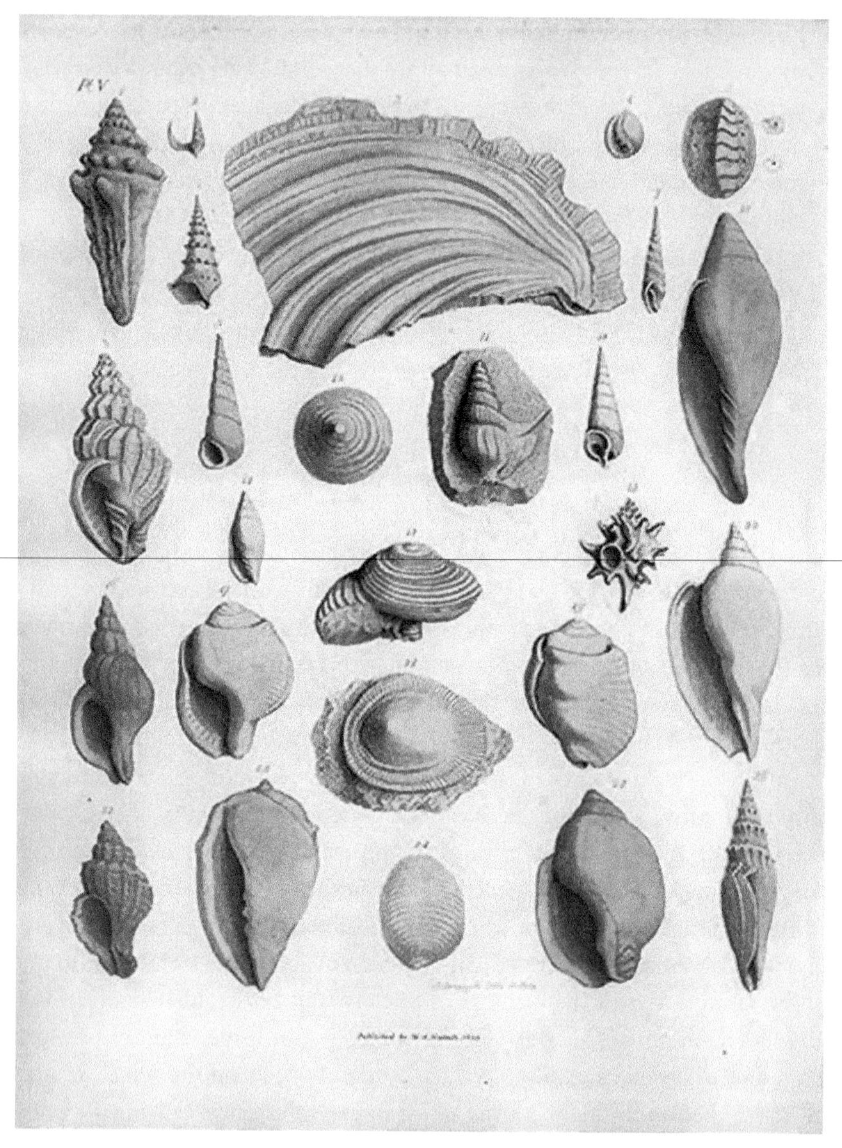

Figure 2.6 An outstandingly beautiful illustration from Parkinson's *Organic Remains of a Former World*. The colouring was often done by his daughter Emma.

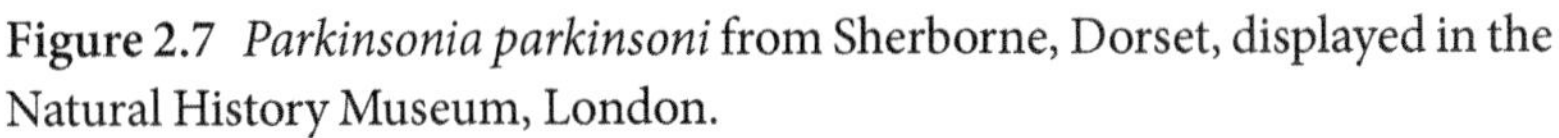

Figure 2.7 *Parkinsonia parkinsoni* from Sherborne, Dorset, displayed in the Natural History Museum, London.
Source: https://commons.wikimedia.org/wiki/File:Parkinsonia_parkinsoni_NHM.jpg.

'He progressed palaeontology from the province of the collector into the realm of real science.'

His knowledge of the geological literature that was available to him is impressive. His prodigious reading in many languages enabled him to synthesise the latest developments in science and reproduce them in an accessible format. However, reconciling his deep belief in the biblical account of the Creation and of Noah's Flood with the fossil record was not easy. In summarising his conclusions at the end of his third volume, he went through the stratigraphic column, providing evidence from each stratum that he considered illustrated that these layers were laid down according to the order stipulated in Genesis. During his career, he appeared to oscillate between accepting and questioning Moses's account of the flood. Moreover, he struggled to explain the evidence of extinctions and the evidence that Earth had a long and complex history, but he believed that the fossil record showed the munificence of his Lord (*Organic Remains of a Former World*, volume 1, p. 8).

> By these medals of creation [fossils] we are taught that innumerable beings have lived, of which not one of the same kind does any longer exist—that immense beds, composed of the spoils of these animals extend for many miles underground, and . . . enormous chains of mountains . . . in which these remains of former ages are entombed . . . are hourly suffering those changes, by which, after thousands of years, they become the chief constituent parts of gems; the limestone which forms the humble cottage of the peasant; or the marble which adorns the splendid palace of the prince. Surrounded, as we are, by the remains of a former world, it is truly surprising, that, in general, so little curiosity and attention are excited by them. Wherever civilized society exists, these wrecks of the earliest ages may be found, yielding to man, the most important benefits.

The struggle to reconcile the account of the Creation and the evidence provided by geology was to occupy geologists and churchmen, including William Buckland, for the rest of the century. As late as 1895, the Reverend Samuel Kinns (1826–1903), whose pious zeal was greater than his scholarship, published a book called *Moses and Geology; or the Harmony of the Bible with Science* and suggested that some of those geologists who doubted the Mosaic account were 'infidels'.

3
The Fossil Collectors

Fossil collectors played a very fundamental role in the study of the geology of the Jurassic Coast. In Lyme Regis, the Anning family ran a fossil hunting and selling business, which advanced science in a way without parallel in the Europe of their day. Here the world's first ichthyosaurs were identified, the world's first complete plesiosaur followed in 1823, and the first British pterodactyl was identified in 1828. This seemed to stimulate the activities of many famous collectors, including the Philpot sisters, who founded the Lyme Regis Museum and put together the collection now in the Oxford University Museum. In 1834, William Buckland arranged for the Swiss palaeontologist Louis Agassiz to visit Lyme Regis to work with Elizabeth Philpot and Mary Anning to obtain and study fish fossils found in the vicinity.

Other significant fossil collectors were Colonel Thomas James Birch (1768–1829), part of whose collection went to Georges Cuvier (1769–1832), the famous French naturalist, and Thomas Hawkins, a maniacal collector of giant fossil reptiles. Another important character was Charmouth quarry owner, James Harrison (1819–1864). He was not a local man by birth, having been born in Purley, Sussex, but came to live in Charmouth with his sisters about 1850, after giving up a career in medicine due to ill health. During the 1850s, while quarrying Black Ven, he discovered the first fossils of a small, four-legged plant-eater, about four to five metres long, called *Scelidosaurus* (see Norman [2021] for a recent discussion of this beast). This was very probably the earliest complete dinosaur fossil to be unearthed in England. Harrison died in 1864, aged only 45 years. Mention also needs to be made of those artists whose illustrations of the fossil bones collected by such figures as Anning and Buckland drew attention to the spectacular finds that were being made. Notable was William Clift (1775–1849), an anatomist of the Royal College of Surgeons, who drew pterodactyl remains for Buckland, and the Reverend George Howland. His daughter married Richard Owen.

The discoveries made by the early collectors formed the basis and inspiration for the world's first palaeoecological reconstruction—*Duria Antiquior* (Ancient Dorsetshire)—which was drawn in 1830 by Henry De la Beche in

Geological Pioneers of the Jurassic Coast. Andrew S. Goudie and Denys Brunsden, Oxford University Press.
 DOI: 10.1093/oso/9780197638088.003.0003

honour of and for the benefit of Anning and her family. *Duria Antiquior* is an imaginative scene, renowned as the first diorama or reconstruction of the environment and animals of a geological period. It is also innovative in that the observer is not out on land, but half in the water. It owes much to collaboration between Buckland and de La Beche.

At the other end of the coast, the vertebrae and toe bones of *Iguanodon* were found before September 1829, in Swanage Bay, weathering out of the Wealden deposits there. These were located by Reverend Thomas Oldfield Bartlett (1788–1841), who was rector of Swanage from 1817 onwards and also vicar of Worth Matravers. Another important early collection was that of Samuel Beckles, who was encouraged by Sir Richard Owen. Stimulated by the collections already made by Charles Willcox and W. R. Brodie, Beckles made a massive excavation in the Purbeck Group strata at Durlston Bay near Swanage and discovered an immensely rich fossil mammal fauna. The finds were collected into a monograph by Owen in 1866, and the collection is now held at the Natural History Museum in London. The beds also contained fossil insects (Westwood 1854). In Weymouth, Robert Damon and his son were also highly successful collectors and exported specimens all over the world (Figure 3.1).

In 1841, the dinosaurs were created—in a taxonomic sense at least. In a landmark paper to the British Association for the Advancement of Science, Richard Owen introduced the term 'Dinosauria' for the very first time. Owen coined the term using a combination of the Greek words *Deinos*, meaning 'fearfully great', and *Sauros*, meaning 'lizard', in order to describe a new and distinct group of giant terrestrial reptiles discovered in the fossil record.

Figure 3.1 *Scelidosaurus harrisonii.*

Source: https://upload.wikimedia.org/wikipedia/commons/1/1f/Scelidosaurus_harrisonii_flipped_transparent.png.

Mary Anning

A Dorset lady with very little formal education and somebody who never wrote any learned papers for scientific journals, Mary Anning was possibly the greatest fossil collector of her age. Hugh Torrens (1995), the great historian of geology, regarded her as 'The Greatest Fossilist the World ever knew.' She has also been described by Larry Davis (2013) as 'Princess of palaeontology and geological lioness'.

Mary (Figure 3.2) was born in Lyme Regis on 21 May 1799. Her parents, Richard and Molly, had no fewer than 10 children, but only two survived—the second of these was Mary. Her father was a cabinet maker and carpenter, a dissenting Congregationalist, and something of a troublemaker. He died in 1810, as a result of consumption and injuries caused by a fall while walking across the cliffs at Black Ven, between Charmouth and Lyme Regis. His premature death left his family destitute. Mary, together with her mother and brother Joseph, who became an upholsterer, made some money by collecting and selling fossils encountered in the Lias (Lower Jurassic) rocks of the

Figure 3.2 A portion of a painting of Mary Anning from a portrait in the Lyme Regis Museum (ASG).

area. Most notable was the discovery and recovery of creatures including the ichthyosaur ('fish-lizard') and the plesiosaur ('near-lizard'). Their first well-known find was in 1811, when Mary was 12 years old; Joseph dug up a four-foot ichthyosaur skull, and a few months later she found the rest of the skeleton. The plesiosaur, a complete skeleton of which was obtained in 1823, was a reptile distinguishable by its small head, long and slender neck, broad turtle-like body, a short tail, and two pairs of large, elongated flippers. One of Mary's specimens is so well preserved that fish bones and scales from its last meal can still be seen inside its ribcage.

In 1820, Lieutenant-Colonel Thomas James Birch (known as Bosvile from 1824), formerly of the Life Guards, organised a London auction sale, including the specimens he had bought from the Annings. The sale, at Bullocks of Picadilly, which attracted buyers from all over Britain as well as Paris and Vienna, raised up to £400. Birch generously donated some of this sum to the Mary family whom, in 1819, he had found on the point of selling their furniture to pay their rent. The publicity this engendered helped secure Mary's reputation as a phenomenally successful and skilled fossil hunter. Further discoveries were to follow. For example, she found evidence that the ink of fossilised squid-like animals had survived, while the winter of 1828 yielded her the first British example of a 'flying dragon', the fossil reptile *Pterodactylus*. *Squaloraja*, a new fossil fish then thought to be an intermediary between sharks and rays, followed in the winter of 1829. She also showed expertise at identifying fossil faeces—coprolites—a theme taken up by William Buckland. When her friend Henry De La Beche painted *Duria Antiquior*, he based it largely on fossils that Mary had found and sold prints of it for her benefit.

The ichthyosaurs, plesiosaurs, and pterosaur that Mary found along with the dinosaur fossils, which were discovered by Gideon Mantell and William Buckland in other parts of England during the same period, showed that during previous eras the Earth was inhabited by creatures very different from those living today and that extinctions of life had occurred. Her contribution had a very significant impact at a time when there was little to challenge the biblical interpretation of the story of Creation and of the Flood, though in France, Cuvier had already produced geological evidence that the simple single flood story was no longer tenable. The remarkable marine reptiles that she unearthed (Figure 3.3) forced the scientific community into looking at different explanations for the causes of changes that had taken place in the natural world.

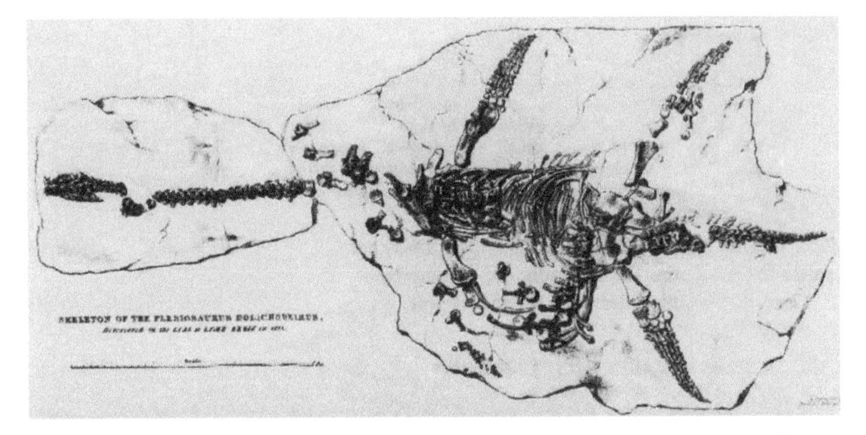

Figure 3.3 Drawing published in the *Transactions of the Geological Society* of the nearly complete *Plesiosaurus dolichodeirus* skeleton found by Mary Anning in 1823.
Source: https://upload.wikimedia.org/wikipedia/commons/e/e4/Anning_plesiosaur_1823.jpg.

Mary did a lot of collecting work with Elizabeth Philpot, who encouraged her to read about geology and understand the science behind the fossils she collected and sold. In 1826, Anning discovered what appeared to be a chamber containing dried ink inside a fossil squid (Figure 3.4). She showed it to Philpot, who was able to revivify the ink by mixing it with water and used it to illustrate some of her own ichthyosaur fossils. Buckland illustrated this in his *Geology and Mineralogy* (1837, plate 19) and in a footnote gave acknowledgement to Anning.

Over the years, Mary's fame spread. In spite of the fact that she was a woman in a man's world and also was a member of a society that was as stratified as the rocks in the Lias, she attracted the attention of some great professional geologists of a far higher social class, including De la Beche, Buckland, and William Conybeare. The Swiss palaeontologist Louis Agassiz visited Lyme Regis in 1834 and worked with Anning to obtain and study fish fossils found in the region. She also was admired by Georges Cuvier in France, who received some of her fossils (Vincent et al. 2014). She even visited London, where she stayed with Sir Roderick Murchison, one of the most patrician of 'professional' geologists. Some of these individuals failed to give her due recognition in their publications, few specimens were named after her, and many collections in museums record the names of the donors but not of the collector herself. However, in her later years, she received some financial

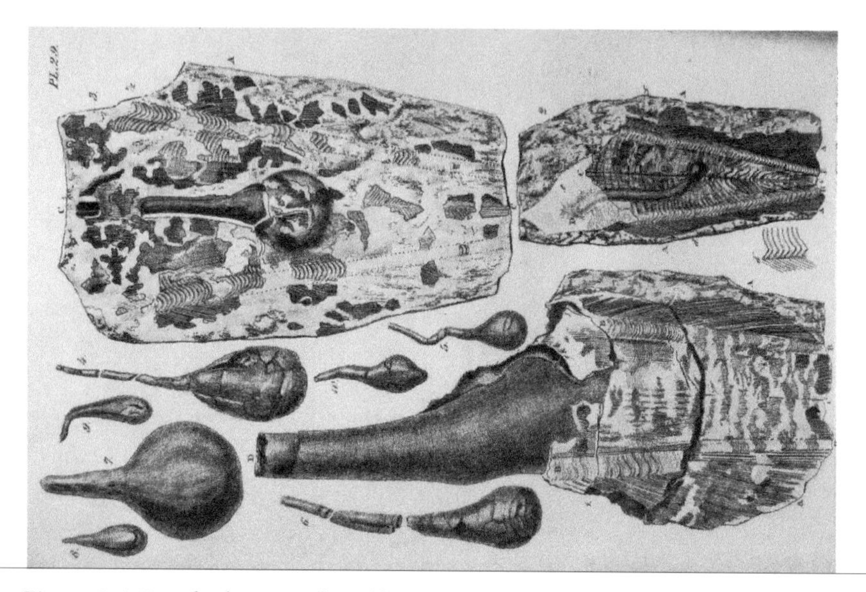

Figure 3.4 Fossil ink sacs as found by Mary Anning and illustrated in Buckland's *Geology and Mineralogy*, 1837.

support from the British Association for the Advancement of Science, from Prime Minister Lord Melbourne, and from members of the Geological Society of London.

In 1845, Anning found that she was suffering from breast cancer, and she died in Lyme Regis on 9 March 1847. She was buried in the churchyard of St Michael's, the local parish church. This is located up the road from what is now the Lyme Regis Museum, a building that is located on the site of the Annings' home and fossil shop (Figure 3.5). Members of the Geological Society contributed to a stained-glass window in her memory, unveiled in 1850. It depicts various acts of mercy: feeding the hungry, giving drink to the thirsty, clothing the naked, sheltering the homeless, and visiting prisoners and the sick. De la Beche paid tribute to her in his Presidential Address to the Geological Society in 1848.

The varying views on her personality are fascinating. The allegedly misogynist dinosaur hunter Gideon Mantell—whose work had been greatly stimulated by Anning's discoveries—described her in 1832 as 'a prim, pedantic, vinegar-looking, thin female, shrewd and rather satirical in her conversation' (see Curwen 1940, p. 108).

Figure 3.5 The plaque outside the Lyme Regis Museum, the location of Mary Anning's house (ASG).

In contrast, she has been characterised by Sir Crispin Tickell (1996) as having 'the sharpest eyes in the business', showing 'immense patience and persistence', being 'physically courageous and tough' (she nearly died in 1833 during a landslide that killed her dog, Tray, and also when young was struck by lightning), and having the ability to put the pieces of a fossil specimen together. Local people regarded her as being 'a kind, straightforward and generous person'. Tickell, a distant descendant, concluded (1996, p. 27) that

> She was no dainty heroine from children's tales, no conventional creature of fantasy, no mere local prodigy, no defender of women' rights, no petrified hand maiden of science. Instead she was a tough, practical, complex, generous, sometimes prickly, independent-minded person of great intelligence, who surmounted the obstacles of her sex and circumstances to help lay the foundations of a new science of the earth.

Elizabeth Philpot

Elizabeth Philpot (1780–1857) was a friend of Mary Anning and a remarkable fossil collector in her own right. She had great knowledge of fossil fish and was known for her extensive collection of specimens. She was consulted by leading geologists and palaeontologists of the time, including William Buckland and Louis Agassiz. As we have already seen, when Anning discovered that the fossil of belemnites contained ink sacs, it was Philpot, a very decent artist, who discovered that the fossilised ink could be revivified with water and used for illustrations.

Philpot and her sisters Louise and Margaret moved from London to Lyme Regis in 1805. They shared a house in Silver Street, now the Mariners Hotel, purchased for them by their brother, a London lawyer. They lived in Lyme Regis for the rest of their lives. Philpot befriended Anning while she was still a child. This was in spite of the fact that there was a 19-year age difference between them and that Anning was from a much poorer social stratum. Whereas Anning collected fossils for sale, the Philpot sisters collected them for their own 'museum'. They were frequently seen collecting fossils together. Philpot never married.

The report on British fossil reptiles published by Owen (1841) has descriptions of *Plesiosaurus* and *Ichthyosaurus* specimens from the Philpot Collection (Figure 3.6), and Buckland's (1829) description of sepia ink from belemnites was based on specimens which had been discovered by Philpot and Anning. Agassiz (1843) made considerable use of specimens from the Philpot Collection in his five-volume *Recherches sur les Poissons Fossiles* and even named the fossil fish *Eugnathus philpotia* in recognition of Philpot's contributions to palaeoichthyology. The Philpot sisters' important fossil collection in 1880 became housed at the Oxford University Museum of Natural History, while the Philpot Museum, now known as the Lyme Regis Museum, was built in Lyme Regis in honour of the sisters by their nephew Thomas Philpot.

Richard Owen

Richard Owen (1804–1892) (Figure 3.7) was a great geologist and anatomist who was noted for three main things: coining the word *Dinosauria* ('terrible

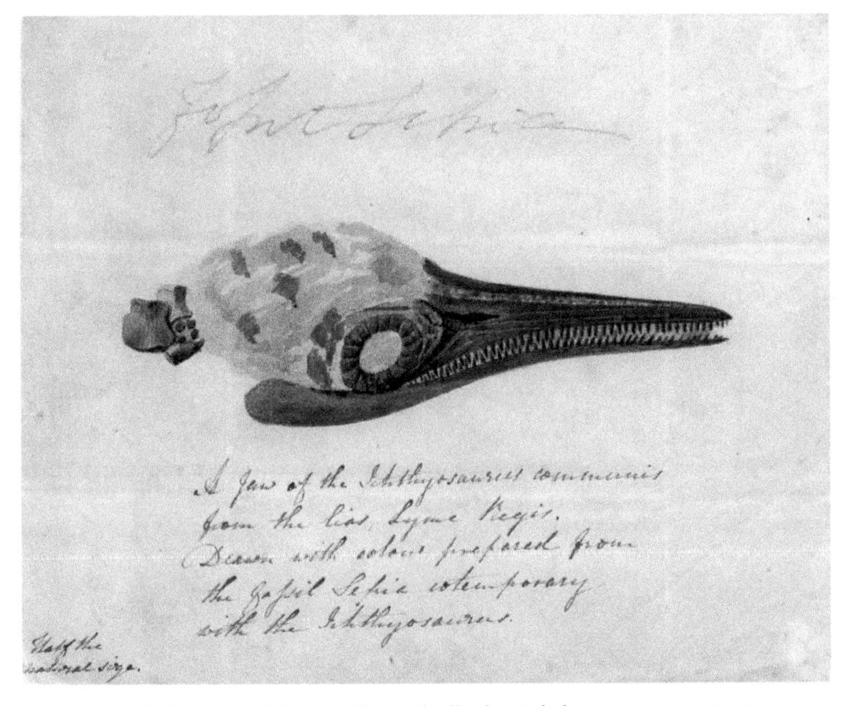

Figure 3.6 A drawing of the fossilised skull of an ichthyosaur, an extinct marine reptile, by Elizabeth Philpot. It was drawn using her own extraordinary technique of revivifying the dried up ink found in fossilised chambers of extinct squid-like creatures called belemnites. The script reads: 'Drawn with colour prepared from the fossil Sepia contemporary with the Ichthyosaurus.'
Source: © Oxford University Museum of Natural History, with permission.

reptile') in 1841 (and publishing it in 1842), helping to establish the Natural History Museum in South Kensington, and making extensive studies of fossil bones. Born in Lancaster, he took up medicine and then became associated with the Royal College of Surgeons. In 1836, he was appointed Hunterian Professor of comparative anatomy and physiology and devoted his life to the study of the anatomy of extinct animals. Owen's role in British science and society of his day was formidable: he had the ear of deans, bishops, and princes; he tutored the children of the Royal Family; and he presided over meetings of learned societies. He lectured the aristocracy and moved freely in their circles despite his middle-class background. The first person to receive an honorary degree from Cambridge, he was a friend of Charles

Figure 3.7 Richard Owen with the bones of a giant extinct moa from New Zealand.
Source: https://en.wikipedia.org/wiki/Richard_Owen#/media/File:Dinornis1387.jpg.

Dickens and the philosopher William Whewell, an advisor to Prime Minister William Gladstone, and an avid cellist with a fine singing voice. He brought dinosaurs to worldwide attention through the remarkable models he helped to create for the Great Exhibition when it reopened in 1854. He was an imposing figure—over six feet tall, thin, and becoming increasingly gaunt with age. He was knighted in 1883.

That said, he was a very controversial figure and was accused of appropriating the work of other scientists as his own, not least that of Gideon Mantell. Some contemporaries described him as deceitful, odious, cantankerous, and malicious. He was famed for feuding with scientific contemporaries, most notably Charles Darwin, and became eclipsed by him. Owen was an enthusiastic Biblical creationist and was intensely critical of 'transmutationism' (evolution). His unwillingness to accept this, despite the growing support for evolution by natural selection after the publication of Darwin's *On the Origin of Species* in 1859, led to a decline in his reputation. In recent years, however, his role has been reappraised, and, to a certain degree, his reputation has been improved (see Rupke 2009). One French scholar described Owen as 'le Cuvier Anglais' (Gaudant 1992).

Indeed, the young Owen had met Cuvier in Paris in 1831. Owen was much intrigued by the specimens that his peers had discovered. He examined William Buckland's *Megalosaurus* specimens at Oxford and Gideon Mantell's *Iguanodon* and *Hylaeosaurus* specimens housed in the British Museum in London. He had also visited Lyme Regis in 1839 with Mary Anning, William Buckland, and William Conybeare. They were almost overwhelmed by the incoming tide. He also saw the collection made by the Philpot sisters.

Not only did Owen coin the term 'dinosaur', but he also used three genera to define and classify his dinosaurs: the carnivorous *Megalosaurus*, the herbivorous *Iguanodon*, and the armored *Hylaeosaurus*. While many of the fossils recovered from the strata of Lyme Regis were those of marine creatures, occasionally some remains of land-living creatures were discovered. One of these was found in 1858, by local quarry owner James Harrison. This dinosaur was called *Scelidosaurus harrisoni* after its discoverer. It was described by Richard Owen in 1861, and was the first essentially complete dinosaur to be discovered and subjected to proper scientific scrutiny. It was also very early, geologically, coming from the Early Jurassic, and was thus quite close in time to the dawn of the dinosaurian era itself.

Samuel Beckles

Samuel Husbands Beckles (1814–1890) (Figure 3.8) was born on the island of Barbados and died in Hastings. He was a Bajan/English nineteenth-century lawyer and, like De La Beche, may have derived some of his wealth from slave

Figure 3.8 Samuel Husbands Beckles.

Source: C. J. Duffin, 2012. https://www.researchgate.net/publication/259884396_Coprolites_and_characters_in_Victorian_Britain/figures?lo=1. Courtesy of Chris Duffield.

ownership. He was the seventh child of John Alleyne Beckles and Elizabeth (née Spooner). His father was at that time a high-ranking judge on the island. Beckles moved to England, becoming a student of the Middle Temple in 1835 and being called to the bar in 1838. He retired from his life as a barrister due to ill health in 1845, moving to St Leonards-on-Sea in East Sussex. Beckles spent his remaining 45 years collecting fossils, apparently unrestricted by his health. He was also a noted poet and art collector.

The Purbeck strata, with their bands of limestone and clay, had long been worked by hand for their building and decorative stones. This provided ample opportunity for the collection of the spectacular and varied fossilised remains of fish and reptiles, along with the occasional plants, abundant molluscs, and other invertebrates that they contained. Coram and Radley (2021) provide a modern analysis of these remarkable beds.

The Reverend J. H. Austen wrote an account of the Purbeck strata in Durlston Bay, which was published in 1852. In 1857, the geologist Hugh Falconer reported on some recently found fossil mammal remains from the Early Cretaceous dirt beds of Durlston Bay in Purbeck. These had been collected in 1854 by the Reverend P. B. Brodie, Mr W. R. Brodie, and Mr Charles Wilcox, a medic from Swanage. Their small collection of small vertebrate remains was sent to Richard Owen at the British Museum. Owen approached Samuel Beckles to undertake a sustained effort to recover more material. Hugh Falconer (1857, p. 261) offered fulsome praise for Beckles.

> Before these remains had reached London, Mr Samuel H Beckles, so favourably known from his researches in Sussex and the Isle of Wight, after free communication with Sir Charles Lyell about the importance of a close and sustained search for mammalian remains at Purbeck, proceeded to Swanage for the express purpose of carrying it out. . . . When the first line of section ceased to be productive, or could no longer be worked, he opened new ground, under difficulties which would have damped the ardour of a less earnest inquirer. . . . So productive have his labours been, that hardly a week passes without its regular instalment of a couple of dispatches of mammalian jaws from Purbeck.

The excavations undertaken by Beckles (Figure 3.9) were very extensive and probably represent one of the largest palaeontological excavations ever undertaken in the United Kingdom (Sweetman et al. 2018). Beckles both supervised and paid for the excavation, which became known as 'Beckles' Pit'. An area of more than 600 square metres was excavated, removing an overburden five metres in thickness. The excavation lasted for nine months and was featured in the *Illustrated London News* (Kingsley 1857).

At least 12 species of mammals were recovered, along with the remains of reptiles, coprolites, insects, and freshwater shells. He discovered the small herbivorous dinosaur *Echinodon*. The only known species, *Echinodon becklesii*, the mammal *Plagiaulax becklesii*, and the dinosaur *Becklespinax* (from Sussex) were named in his honour. In 1862, he wrote a paper about the existence of dinosaur footprints from Swanage. In 1859, he was elected as a Fellow of the Royal Society in recognition of his ability and enthusiasm in his palaeontological endeavours.

Figure 3.9 The excavations at Durlston Bay. The gentleman with the top hat is probably Samuel Beckles himself.
Source: https://upload.wikimedia.org/wikipedia/commons/5/5f/Beccles_Pit_ILN_1857.jpg.

Peter Brodie

Peter Bellinger Brodie (1815–1897) was born in London and, like so many contemporaries, was both a geologist and churchman. While residing with his father, a noted lawyer, at Lincoln's Inn Fields, he gained an interest in fossils from visits to the museum of the Royal College of Surgeons at a time when William Clift (who was father-in-law to Richard Owen) was its curator. Through the influence of Clift, he was elected a Fellow of the Geological Society of London in 1834. Proceeding to Emmanuel College, Cambridge, Brodie came under the influence of the great Adam Sedgwick and helped him in the Woodwardian Museum. He then devoted his time to geology and the church. Entering the ministry in 1838, he was curate at Wylye in Wiltshire and, for a short time, at Steeple Claydon in Buckinghamshire. Later, he became rector of Down Hatherley near Cheltenham in Gloucestershire and, finally, in 1855, vicar of Rowington in Warwickshire. In 1887, the Murchison

Figure 3.10 The Reverend Peter Berrington Brodie.
Source: H. B. Woodward (1897).

Medal was awarded to him by the Geological Society. He died at Rowington in 1897. His insect collections are mostly in the Natural History Museum, London. Others are in the collections of the British Geological Survey at Keyworth, Warwick County Museum, Dorset County Museum, Bath Royal Literary and Scientific Institution, and Gloucester City Museum.

Brodie (Figure 3.10) is important because fossil insects formed the subject of his special studies (Brodie 1845), and many of his published papers relate to them (Figure 3.11). He was more or less the first English palaeontologist to develop this field of study, which is now called palaentomology. While other were pursuing huge dinosaurs, Brodie was pursuing small bugs. As he wrote (p. 113),

> One object in the present imperfect sketch has been to show the value and importance of even minute investigations, in elucidating the conditions under which certain strata were deposited, and more especially to point

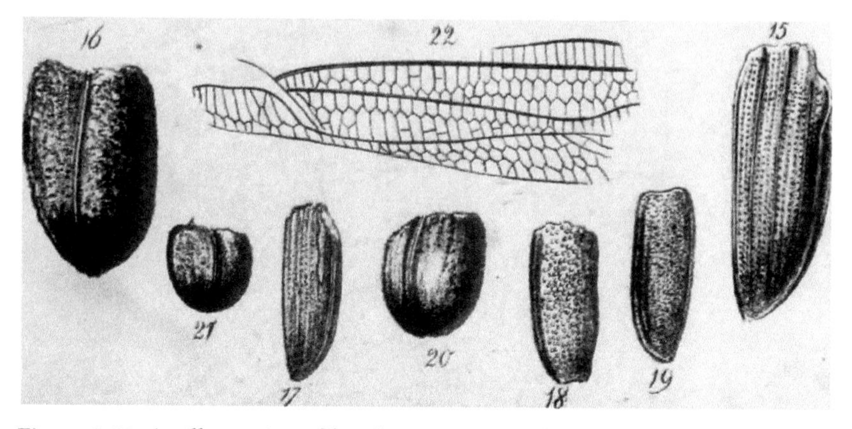

Figure 3.11 An illustration of fossil insect remains from Brodie's book of 1845.

> out the existence of many of those fragile but beautiful forms of animal life which tenanted our earth at very early geological periods. Our minds are so constituted that we readily admire everything grand or sublime in nature; but we are apt to overlook those small and less striking objects, which are, in fact, equally worthy of our admiration and regard.

He also believed (p. 114) that the fossil insects showed the munificence of our Creator.

> Thus, then, in all our labours and reasonings it should never be forgotten, in any case, that every examination into the wonders and beauties of God's creation, whilst it increases our knowledge and improves the understanding, has also a far higher and better purpose in displaying the glory of God, and in leading us to adore and praise the wisdom and omnipotence which are daily displayed in the material world.

Brodie had relatives in Swanage, including W. R. Brodie, a cousin, and in the 1850s undertook work on the fossil insects found in that area. He collected them from the sites that had been investigated by Samuel Beckles and also from the Purbeck beds elsewhere, from some Tertiary clays and from Ringstead. Brodie collaborated with John Westwood, Professor of Zoology at Oxford University, on these finds. Work on the rich insect fauna from Durlston, made famous by Brodie's work, continues to this day (see Coram and Jepson 2012).

Westwood (1854, p. 379) remarked of Brodie that he 'has been highly successful in detecting minute fragments of insect remains in small slabs of stone, which would to a less educated eye have been passed over as destitute of any traces of ancient animal life'.

Thomas Hawkins

Thomas Hawkins (1810–1889) (Figure 3.12), a fossil collector, liked big beasts. He was born on 25 July 1810 near Glastonbury. From about 1832 to about 1845, Hawkins lived at Sharpham Park, also near Glastonbury. He claimed to have formed the ambition at the age of 18 to amass the greatest collection of fossils in the world, and he succeeded in this thanks both to huge determination and to his skill as a fossil restorer. Unfortunately, he touched up imperfect specimens with plaster, sometimes adding whole

Figure 3.12 Thomas Hawkins.
Source: https://commons.wikimedia.org/wiki/File:Thomas_Hawkins,_geologist_(1810-1889).jpg.

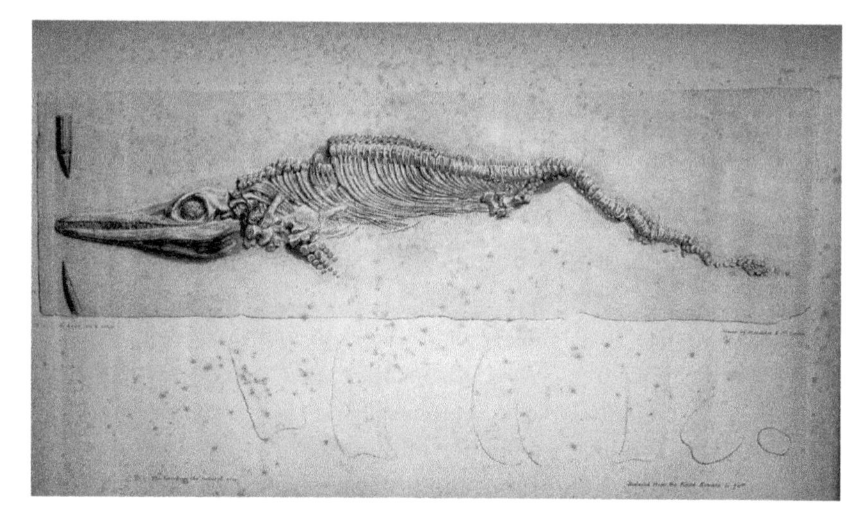

Figure 3.13 One of Hawkins's splendid fossils.
Source: https://commons.wikimedia.org/wiki/File:HawkinsSeaDragonsPlate2.jpg.

limbs of his creation. As Christopher McGowan remarked (2001, p. 130), 'The end product often looked more perfect than nature intended.' Whether such actions were the result of overenthusiasm or fraudulence has been debated, but the results remain among the finest exhibits in the Natural History Museum in London. Hawkins also took great liberties with zoological nomenclature, inventing new names for species which had already received their scientific names.

A Fellow of the Geological Society of London from the tender age of 22, he collected superb dolphin-like ichthyosaurs and long-necked plesiosaurs from Jurassic rocks (Figure 3.13) through quarry men at Street, Somerset, and through professional collectors such as Mary Anning of Lyme Regis. He sold two collections to the British Museum (1834 and 1840) for a massive £3110 5*s* (about £300,000 in today's money). Together with Gideon Mantell's collection, these made the museum the preeminent collection of fossil reptiles. Hawkins also donated smaller collections to the universities of Cambridge (1856) and Oxford (1874). Those in Oxford still hold a prominent position on the rear wall in the great hall of the University Museum of Natural History. His collections contain some of the oldest known complete plesiosaurs, and they remain scientifically important. Two huge, but weird books (both dedicated to Buckland), *Memoirs on Ichthyosauri and Plesiosauri* (1834) and *The Great Sea Dragons* (1840), mixed anatomical description and

Figure 3.14 The frontispiece of Hawkins's book *The Great Sea Dragons*. The creator was John Martin (1789–1854). It depicts Jurassic life in the artist's typical nightmarish and gothic style. On the left, two plesiosaurs are attacking an ichthyosaur, whilst on the right pterosaurs are scavenging the corpse of another. It dates from 1840.

Source: https://en.wikipedia.org/wiki/Thomas_Hawkins_(geologist)#/media/File:Great_Sea-Dragons.jpg.

fine illustration (Figure 3.14) with autobiography, sacred history, theological discussions, and grandiloquent visions of the 'Gedolim Taninim of Moses, extinct monsters of the Ancient Earth'.

Hawkins was a social failure, presumably due to his eccentricity. While he himself mentions his own deafness and tinnitus, the underlying problem was evidently a serious personality disorder of unclear nature and onset. He seems to have been unpleasant, litigious, difficult, and very possibly profoundly disturbed, even delusive. To equals and inferiors Hawkins was dangerously quarrelsome. Somerset people remembered him for feuds and provoking riots. He was said to be 'very near the border between eccentricity and criminal insanity' (Taylor 2003). He had no known profession or trade other than his sale of specimens, and, as far as is known, he lived off this capital supplemented with inheritances and a marriage settlement. He died at Ventnor, Isle of Wight, on 15 October 1889 and was buried there under a grandiloquent epitaph bordered by two great granite columns.

Robert Damon

Robert Damon (1814–1889) wrote a classic book *Geology of Weymouth and the Isle of Portland; with notes on the Natural History of the Coast and Neighbourhood* (first edition in 1860 and the second edition in 1884). He was also a major dealer in geological specimens and provided barnacles to Charles Darwin. He was admired by Horace Woodward. He also seems to have travelled and to have obtained specimens from Lebanon, Pompeii, and Russia.

Of Flemish descent, he was born in Melcombe Regis, Weymouth, in 1814, and initially was a hosier and glover. He then established his dealership and sold natural history specimens throughout the world, including to museums in Australia and New Zealand. He ran a fossil shop at Augusta Place, the Esplanade, Weymouth. This is now a fish and chip shop (Figure 3.15). His business was continued by his son, Robert Ferris Damon (1845–1929), who was particularly appreciated for the casts he made of anthropological specimens. However, Robert senior developed scientific interests and was especially interested in shells. His importance was shown by his election as a Fellow of the Geological Society of London, by his election as a member of the Imperial Natural History Society of Moscow, and by the fact that he

Figure 3.15 The site of the Damon shop in Augusta Place, Weymouth (ASG).

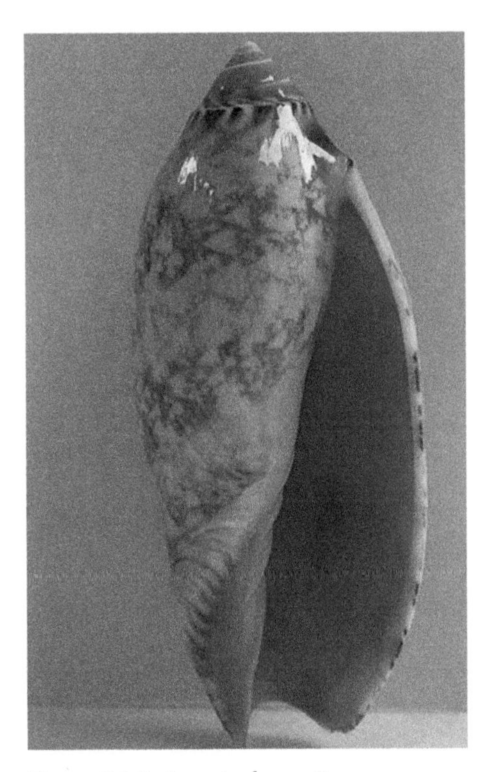

Figure 3.16 *Amoria damonii.*
Source: https://en.wikipedia.org/wiki/Amoria_damonii#/media/File:Amoria_damonii_001.jpg.

had two species named in his honour: *Amoria damonii* (Figure 3.16) and *Paramelania damoni* (Figure 3.17). The former is a marine gastropod, while the latter is a species of tropical freshwater snail.

Damon's book on the local geology is a concise and handy guide, and he acknowledges his debt to H. W. Bristow, the indefatigable mapper from the British Geological Survey, who had produced a vertical section of the rocks between Portland and Somerset. Damon discusses all the major geological formations, from the Inferior Oolite of the Jurassic through to the Pleistocene. While much of the book is based on a very comprehensive analysis of the available literature, some of his sections were used by later writers, including W. J. Arkell (1947), and he also includes many valuable observations of his own on such phenomena as cave deposits, the burning cliff in Ringstead Bay, mammoth remains, dene holes (medieval chalk extraction pits), and landslips. His treatment of the origin of Chesil Beach is

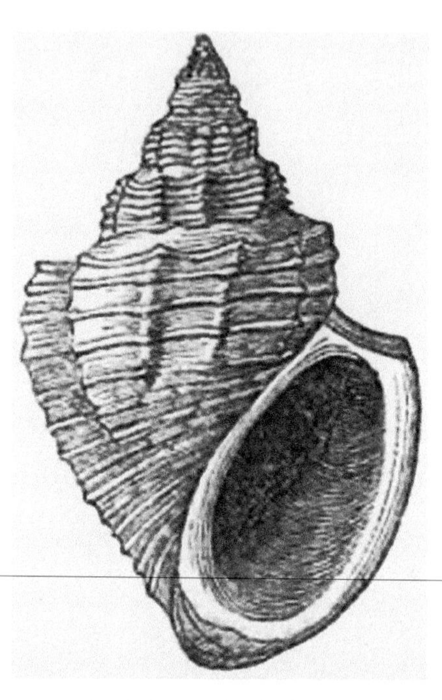

Figure 3.17 *Paramelania damoni.*
Source: https://en.wikipedia.org/wiki/Paramelania_damoni.

both sensible and largely right, and he was struck by its importance (1884, p. 167).

> The Chesil Bank, situated within one of the largest bays in the kingdom, is the most remarkable formation of its kind. Its magnitude exceeds that of any other in Europe, and it may fairly rank in interest with the mountain chains and other monuments of physical geography.

Equally, he was an acute observer, and his discussion of a landslip on Portland (1884, p. 157) demonstrates this. He remarked that on Portland landslipping

> . . . is attributable to the position of the Kimmeridge Clay, which forms the substratum of the Island. The water passes through the numerous cracks which traverse the solid beds above and behind, and causes an undermining and consequent precipitation of the superincumbent mass. This was illustrated in the landslip of Dec 26, 1858, when the sliding down of an

> extent of undercliff covering an area of twenty-five acres caused the sinking of an enormous mass of broken stone, the debris of adjoining quarries, and the accumulation of very many years. The scene of this occurrence was on the west side of the Island overlooking the West Bay. In ascending from the beach shortly after the occurrence, the observer would be first attracted by the low undercliff of Kimmeridge Clay, which from lateral pressure, was pushed forward beyond the beach into the sea, and forced upwards with the shingle over it, so as to present a steep outer face towards the sea. A little way up the cliff a singular change was effected in the position of some garden plots, which previously inclined towards the sea at an angle of 45°, but dip now as much in the opposite direction, the plane of this portion of land having consequently traversed 90°, or one-fourth of the circle.

This is an early observation of what is now called a *rotational slip*. Damon was an amateur and untrained geologist, but one of great energy and perspicacity. He died in Weymouth in 1889, aged 75 years.

4

Establishment of Geology at the Universities and the Golden Age

The lectures of John Kidd (1775–1851), who was Professor of Chemistry at Oxford University during the years 1805–1810, introduced into the University for the first time the subject of geology, in which he took a keen interest. In his 'subterranean class-room' in the old Ashmolean Museum, he taught figures such as William Buckland and William Conybeare. When Kidd resigned the Readership of Mineralogy in 1813, he was succeeded by Buckland.

In Cambridge, Adam Sedgwick (1785–1873), who had been a mathematics tutor, was appointed Woodwardian Professor in 1818. Both Buckland and Sedgwick were keen and eloquent teachers and raised the interest and zeal of many who, in after years, became distinguished geologists (Woodward 1911, pp. 46–50). At University College London, geology was at first offered in a series of subscription lectures, presented in 1830 as one part of the courses in chemistry, botany, and zoology by John Phillips (1800–1874). It was in 1841 that the first professor, Thomas Webster, who worked on the Jurassic Coast, was appointed and courses provided on a regular basis. At the newly formed King's College, London, Charles Lyell (1797–1875) was Professor of Geology from 1831–1833. Lyell's short period at King's was explained thus by Sir Edward Bailey (1962, p. 98):

> Lyell's first course of lectures was delivered in 1832, and his last in 1833. To begin with he was able to throw his lectures open to the public, including ladies: but before long the college authorities excluded the latter, because their presence 'diverted the attention of the young students'. Attendance dropped precipitously, and Lyell resigned. He felt that his proper place in life was that of gentleman-scientist-author, without strings.

Geological Pioneers of the Jurassic Coast. Andrew S. Goudie and Denys Brunsden, Oxford University Press.
 DOI: 10.1093/oso/9780197638088.003.0004

Nonetheless, King's produced some great geologists, including Henry William Bristow, Edward Forbes, and Osmond Fisher, all three of whom worked on the Jurassic Coast.

Horace Woodward (1911, p. 58), in his short *History of Geology*, described the period from *c.* 1820 to 1840 as being the 'Golden Age of Geology'. It was the time of 'the Great Masters of Geology: a period characterised by strenuous field-work and by many grand discoveries in various parts of Europe and America'. In Britain it was the era when William Buckland, Roderick Murchison, Adam Sedgwick, Charles Lyell, and Henry De la Beche were in their prime. Some of these masters played a major role in deciphering the geological history of the Jurassic Coast.

Also in the early nineteenth century, the study of geology started to become more professional. On 13 November 1807, a group of geologists met in the now demolished Freemasons Tavern in Great Queen Street, Covent Garden, and established the Geological Society of London. Rather later, there was the establishment of the Geological Survey, and this is associated with the name of Henry De la Beche. Later still, the Geologists' Association was established in 1858. It was established because, in August of that year, a letter appeared in the magazine *The Geologist* proposing the formation of 'an Association of Amateur Geologists' so that 'solitary' students of the science could form a society where they 'could compare notes, give an account of our rambles, examine one another's fossils and minerals and . . . be of great assistance to one another'. In the formation of the Association, particular emphasis was placed on the holding of 'Excursions or Field Meetings', and the reports of these have been a fertile source of information on key sections and geological phenomena, not least along the Jurassic Coast.

It is important to see these developments against the religious background of the time, for society and individual geologists were often devout members of the Anglican Church. They believed in the biblical story of the Creation and in the account of Noah's Flood, though some, like James Hutton, had already produced field observations that countered this literal interpretation. They also thought that the Earth was only around 6000 years old. Their ideas occurred before Darwin had produced his theory of evolution. Given this framework of ideas, it was difficult to account for the evidence that was emerging in the geological record of massive changes in the environments of rock deposition and of multiple phases of the extinction and creation of organisms. Geologists invoked various catastrophes, including the Flood, to account for this seemingly complex history. It was only in the first decades of the nineteenth century that this approach, called 'catastrophism', was gradually replaced by the idea that the Earth had a long history and had been moulded by many of the processes we see in operation today, an approach called 'uniformitarianism' or 'gradualism'. Moreover, it was only from the 1820s that geologists began to recognise that there had been an Ice Age in which various phenomena, including valleys, lakes, and spreads of 'drift', had been created.

William Buckland

William Buckland (1784–1856) (Figure 4.1) was an English theologian who became Dean of Westminster. However, he was also a geologist and palaeontologist who made many contributions to our knowledge of the Jurassic Coast. He was born in Axminster, Devon, which is located just 10 kilometres inland. As a child he would accompany his father, Charles, the rector of Templeton and Trusham, on his walks, collecting fossil shells, including ammonites from the Lias rocks exposed in local quarries. He was educated at Blundell's School in Tiverton, Devon, before going to Oxford University, where he graduated in 1805 at Corpus Christi College. During the vacations he paid frequent visits to Lyme Regis, where he first met Henry De la Beche and Mary Anning. In the summer of 1808, Buckland made his first geological tour, alone on horseback, from Oxford across the chalk downs to Corfe Castle in the Isle of Purbeck.

He became Reader in Mineralogy in 1818, and he gave lectures in geology in Oxford for many years (Figure 4.2). He had a reputation as a somewhat unorthodox teacher (http://www.oum.ox.ac.uk/learning/pdfs/buckland.pdf; p. 2, accessed 3 June 2020).

> He paced like a Franciscan preacher up and down behind a long showcase. . . . He had in his hand a huge hyena's skull. He suddenly dashed down the steps—rushed skull in hand at the first undergraduate on the front bench and shouted 'What rules the world?' The youth, terrified, answered not a word. He rushed then on to me, pointing the hyena full in my face—'What rules the world?' 'Haven't an idea', I said. 'The stomach, sir!', he cried 'rules the world. The great ones eat the less, the less the lesser still!'

Figure 4.1 William Buckland, President of the Geological Society of London.
Source: https://en.wikipedia.org/wiki/William_Buckland#/media/File:William_Buckland_GSL.jpg.

In addition, his sense of humour was said to be coarse. To earnest Victorians such as Charles Darwin, his lectures and contributions to debates at the Geological Society of London were seen as undignified buffoonery. However, they went down very well with Oxford students.

In 1825, he married Mary Morland, who had a splendid reputation as an illustrator of fossils. Together they had a large family—nine children, five of whom survived into adulthood. Their accommodation in Christ Church,

Figure 4.2 Buckland lecturing in the old Ashmolean Museum, Oxford, in 1823, with visual aids. Coloured lithograph printed by C. Hullmandel after N. Whittock.

Source: Wellcome Collection. https://wellcomecollection.org/works/zy29gh2t?wellcomeImagesUrl=/indexplus/image/V0006732.html.

Oxford, where he became a Canon, was amazing. As Noel Annan recounts (1999, p. 7),

> When you entered the hall you might as easily mount a stuffed hippopotamus as the children's rocking horse. Monsters of different eras glared down on you from the walls. The sideboard in the dining room groaned under the weight of fossils and was protected from the children by a notice: PAWS OFF. The very candlesticks were carved out of the bones of Saurians. Toads were immured in pots to see how long they could survive without food. There were cages full of snakes, and a pony with three children up would career round the dining-room table and out into the Quad. Guinea pigs, owls, jackdaws and smaller fry had the run of the house. The children imbibed science with their mother's milk. One day a clergyman excitedly

> brought Buckland some fossils for identification. 'What are these, Frankie?' said the professor to his four-year-old son. 'They are the vertebuae of an ichthyosauwus', lisped the child. The parson retired crestfallen to his parish.

One of Buckland's many eccentricities was his passion for zoöphagy (feeding on animals). No living creature was secure from his mania. He claimed to have eaten his way through the animal kingdom, and at his soirées served dishes than included mice on toast, puppies, panthers, moles, crocodiles, and bluebottles. He is even alleged to have gobbled down the heart of Louis XIV of France, which had been preserved in a silver casket at Nuneham, near Oxford.

Buckland did a great deal of work on remains found in caves and fissures. In Goat's Hole Cave, in South Wales, he discovered a human skeleton, which he named the Red Lady of Paviland. He believed it to be the bones of a local prostitute, possibly dating from Roman times, but the reality is that it was the skeleton of a man dating from more than 30,000 years ago. He also worked on Kirkdale Cave in Yorkshire, discovered in 1821, which he believed was an antediluvial den of hyenas. Buckland determined that the bones were from the remains of animals brought into the cave by hyenas who had been using it for a den and not a result of the biblical Flood floating animal remains in from distant lands as he had first thought.

For all his eccentricities, Buckland was a remarkable geologist. One of his early achievements was to establish the existence of great fossil reptiles, the bones of which he found at Stonesfield, a village in the Cotswolds. These enormous bones were of staggering importance, and, in 1824, he announced their discovery to the Geological Society of London. He declared that they were the bones of a giant reptile, which he named *Megalosaurus*, or 'great lizard' (Figure 4.3). This was the first paper to give a full account of what would later be called a dinosaur.

A phenomenon associated with the great fossil reptiles studied by Buckland were what he termed 'coprolites' (fossil faeces). Mary Anning had noticed that stony objects known as 'bezoar stones' (calcareous balls formed in the stomachs of oriental goats and used extensively in medieval times for medical treatment) were often found in the abdominal region of ichthyosaur skeletons found in the Lias at Lyme Regis. She also noted that if such stones were cracked open, they often contained fossilised fish scales and bones, some of which came from small ichthyosaurs. These observations

Figure 4.3 A reconstruction of *Megalosaurus bucklandii* from the Oxford University Museum of Natural History. Artwork by Julius T. Csotonyi.
Source: https://www.oumnh.ox.ac.uk/megalosaurus-and-oxfordshire-dinosaurs. © Oxford University Museum of Natural History.

led Buckland to propose, in 1829, that the stones were fossilised faeces. The name is derived from the Greek words κόπρος (*kopros*, meaning 'dung') and λίθος (*lithos*, meaning 'stone'). Once Buckland realised that coprolites could reveal the diet of the animals that produced them, it opened up a whole new way to investigate the fauna and flora of the past. He recognised that coprolite analysis could reveal if an animal were an herbivore, a carnivore, or both and, in many cases, could determine the species composition of the meal that the animal had eaten. He was able to establish the food chain of the reptiles. This, along with Anning's fossil collections, resulted in the first pictorial representation of a scene from the past painted by his colleague Henry De la Beche in a wonderful composition called *Duria Antiquior*. Spot the coprolites being deposited (Figure 4.4).

Buckland was so fascinated with the beauty of coprolites (Figure 4.5) and what they could tell us about past life that he had a table with thin-sliced fish coprolites embedded in its surface. It is rumoured that he had guests dine at the table prior to telling them about the origins of the rock upon which they had just eaten. The table is in the Lyme Regis Museum. The best collection of Buckland's coprolites is in the Oxford University Museum.

Buckland, in addition to being an Oxford academic, was also a devout Anglican clergyman. In the first decades of the nineteenth century, it was widely accepted that much erosion and the excavation of valleys had been

Figure 4.4 *Duria Antiquior* (Ancient Dorset).

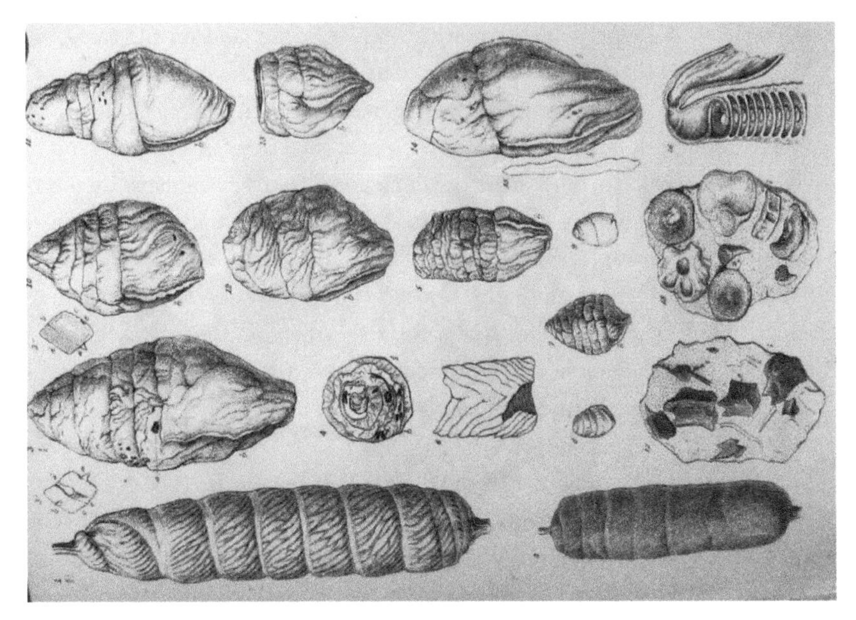

Figure 4.5 Coprolites from Lyme.
Source: W. Buckland, *Geology and Mineralogy* (1837).

Figure 4.6 Buckland's view of the diluvian valleys of the Devon Coast, published in 1822.

achieved by Noah's Flood (Figure 4.6). Buckland embraced this idea, partly on account of his studies on the valleys of south Devon and Dorset, which he believed had a diluvial (i.e. Flood) origin. He first published his arguments in 1822, and then in 1823 he produced his great work *Reliquiae Diluvianae* (p. 239).

> Some of the best examples I am acquainted with of valleys thus produced exclusively by diluvial denudation occur in those parts of the coasts of Dorset and Devon which lie on the east of Lyme, and on the east of Sidmouth; and the annexed views and map will illustrate, better than any description, the point I am endeavouring to establish. In passing along this coast . . . we cross, nearly at right angles, a continual succession of hills and valleys, the southern extremities of which are abruptly terminated by the sea; the valleys gradually sloping into it, and the beach or undercliff, with a perpendicular precipice. The main direction of the greater number of these valleys is from north to south; that is, nearly in the direction of the strata in which they are excavated: the streams and rivers that flow through them are short and inconsiderable, and incompetent, even when flooded, to move anything more weighty than mud and sand.

But, later in his career, Buckland believed that the glacial theory of Louis Agassiz, which established the existence of a great Ice Age, gave a better explanation of most phenomena than Noah's Flood and played a significant role in promoting it. In his famous *Bridgewater Treatise* (Buckland 1837), he acknowledged that the biblical account of Noah's Flood could not be confirmed using geological evidence. By 1840, he was actively promoting the view that what had been interpreted as evidence of the 'universal deluge' two decades earlier was in fact evidence of a major glaciation.

In 1835, along with De la Beche, Buckland published a major paper, 'On the Geology of the Neighbourhood of Weymouth and the Adjacent Parts of the Coast of Dorset'. This is a masterful and well-illustrated summary of the geology of the area and also draws attention to such features as the widespread solution of the Chalk, the importance of faults, and the formation of the sarsen blockstream of the Valley of Stones at Portesham. He also recognised in the Purbeck Beds of Portland that there was a fossil soil layer—the great dirt bed—that contained some fossil species, including cycads, which hinted at the environment—a tropical climate—in which these materials were deposited.

Landslips are a recurring feature of the Jurassic Coast. One of the most spectacular occurred on Christmas Day 1839, to the west of Lyme Regis, in the area now known as the Undercliff. This slip affected a large tract of land below Bindon Manor and Dowlands Farm, resulting in the features called Goat Island and the Chasm. This particular landslip was very well documented because Buckland and his wife were staying with the Reverend William Conybeare and his family. Although Conybeare was a local clergyman, he was also a great geologist. In 1840, they produced what is quite possibly the first detailed scientific report about a major landslip, which was beautifully illustrated with coloured maps and diagrams by William Dawson and engravings by Mary Buckland. Remarkably, the geotechnical mechanism of the landslips, involving a translational slide, has recently been recognised as essentially correct by Ramues Gallois (2010), although this is still the subject of intensive investigation as new mapping takes place (see also, Edmonds 2019).

In 1845, Buckland, who became a recipient of the Geological Society's Wollaston Medal in 1848, was appointed as Dean of Westminster, which carried with it the rectorship of Islip near Oxford (see Box 4.1). He kept a menagerie at the rectory, including the bear Tiglath Pileser, which roamed around the village, visiting the local shop in search of sweets. Initially he approached this new career with great energy, but towards the end of 1849, he fell ill, and suffering from depression, apathy and probable dementia, he spent time in John Bush's Mental Asylum at Clapham in London. He died in 1856. Fittingly, he was buried in the Jurassic rocks of Islip but has a memorial plaque in Westminster Abbey.

Box 4.1 Award Winners

Some Jurassic Coast Recipients of the Wollaston Medal

The Wollaston Medal is the Geological Society's premier award and, in the words of Simon Winchester, 'the Oscar of the world of rocks' (Winchester, S. 2001, *The Map that Changed the World*. Viking, London).

1833 Richard Owen
1844 William Conybeare
1848 William Buckland
1849 Joseph Prestwich
1852 William Henry Fitton
1855 Henry De la Beche
1873 Philip de Malpas Grey Egerton
1909 Horace Woodward
1913 Osmond Fisher
1919 Aubrey Strahan

Some Jurassic Coast Recipients of the Geological Society's Murchison Award

1887 Peter Bellinger Brodie
1893 Osmond Fisher
1897 Horace Woodward
1901 Alfred Jukes-Browne
1914 William Ussher
1991 Michael House

Some Jurassic Coast Recipients of the Lyell Medal

1913 Sydney Buckman
1928 William Lang
1945 Leonard Spath
1949 William Arkell

Mary Buckland

Two pioneers of geology on the Jurassic Coast have been associated with the brewing industry. One of these was William Joscelyn Arkell, and the other was Mary Morland Buckland (1797–1857) (Figure 4.7). She was born in Abingdon, then situated in Berkshire, the eldest daughter of Benjamin Morland (1768–1833), a member of the West Ilsley brewing family, and his wife, Harriet Baster (1777–1799).

Buckland became a palaeontologist and scientific illustrator. Geology was part of her early life. She was encouraged by her father, but when he remarried, she then lived with Sir Christopher Pegge's family—Sir Christopher (1765–1822) was the Regius Professor of Anatomy at Oxford who also taught geology and mineralogy. She inherited his fossil collections following his death.

As a young woman Mary became involved with the new science of palaeontology, drawing beautiful illustrations of fossils for notable scholars such as William Conybeare. She also worked on fossil curation at the Oxford University Museum, undertaking tasks such as labelling and conservation. In addition, she collected fossils, and Louis Agassiz and Charles Lyell described rare specimens of sponges she found.

Figure 4.7 Mary Buckland.

Source: https://en.wikipedia.org/wiki/File:Mary_Morland_Buckland,_from_an_original_photograph.jpg.

While still a teenager, Mary corresponded with the Frenchman Georges Cuvier, then at the peak of his career as the effective founder of palaeontology. She sent him specimens and produced illustrations for him, and, around 1817, he sent her a copy of his latest book *Le Règne Animal Distribué d'après son Organisation: Les Mammifères et les Oiseaux.*

The story, which sounds plausible, goes that she was reading this on a coach journey to Dorset and got into conversation with a fellow passenger who was reading the same book. That passenger was William Buckland, Professor of Geology and Mineralogy at Oxford University, who was 13 years her senior. On 31 December 1825, at Marcham, then in Berkshire, they were married. Buckland played a crucial role in the education of the five of their nine children who survived childhood: the eldest, Frank (1826–1880), became a renowned naturalist and continued the family tradition of zoöphagy. He has become known as 'the man who ate the zoo'.

Besides her child-rearing duties and public service (she supported poor Jewish families in Oxford, for example) Mary played a vital, though largely now forgotten, role in her husband's success as geologist and founder of the Oxford geology department (Figure 4.8). She accompanied him on a year-long honeymoon tour of the continent. She corrected his fine prose and wrote much of it at his dictation. Her skill as an artist was also put to use. She illustrated *Reliquiae Diluvianae* (1823) and his Bridgewater Treatise, *Geology and Mineralogy* (1837). She was also adept at conserving fossils with specially developed cements, in making models of them, and in assisting William's experiments to reproduce fossil tracks—all vital evidence for his scientific endeavours.

Mary died on 30 November 1857, at St Leonards, Sussex. She was buried next to her husband in the graveyard of the parish church at Islip, Oxfordshire (Figure 4.9).

William Conybeare

William Conybeare (1787–1857) (Figure 4.10) was a geologist, palaeontologist, and clergyman. He is probably best known for his groundbreaking work on marine reptile fossils in the 1820s, including important papers for the Geological Society of London on ichthyosaur anatomy and the first published scientific description of a plesiosaur.

Figure 4.8 Silhouette of William Buckland and his wife Mary, both examining their respective palaeontology collections. Their son Frank is playing underneath the table. Taken from the 'Life and Correspondence of William Buckland, DD, FRS, Sometime Dean of Westminster, Twice President of the Geological Society and First President of the British Association', by Elizabeth Gordon, John Murray, London (1894, p. 103).

He came from a race of clergymen, being a grandson of John Conybeare, Bishop of Bristol (1692–1755), and son of Dr William Conybeare, rector of Bishopsgate. Born in London, he was educated at Westminster School. In 1805, he went to Christ Church, Oxford, where, in 1808, he took his BA degree, achieving a first in classics and a second in mathematics. However, while in Oxford he was attracted to the study of geology by the lectures of Dr John Kidd. He was ordained deacon in 1813, was married in 1814, and left the University. He settled as a curate in Suffolk.

In 1815, Conybeare's father died. His father had 'received, for thirty to forty years, an annual income of not less than £3500 from ecclesiastical

Figure 4.9 The combined granite grave of William and Mary Buckland at the parish church at Islip (ASG).

Figure 4.10 William Conybeare.

Source: https://en.wikipedia.org/wiki/William_Conybeare_(geologist)#/media/File:William_Conybeare.gif.

preferments alone' (Torrens 2004), so his death meant that Conybeare was to have few financial worries for the remainder of his life. Conybeare then made extended journeys in Britain and on the continent, some with William Buckland, and he became one of the early members of the Geological Society of London. Both Buckland and Adam Sedgwick, two fellow Anglican clergymen, acknowledged their indebtedness to him for instruction received when they first began to devote their attention to geology. In 1821, in collaboration with Henry De la Beche, he distinguished himself by describing, from fragmentary remains, the saurian *Plesiosaurus* in a paper for the Geological Society that also contained an important description and analysis of all that had been learned to that point about the anatomy of ichthyosaurs (Figure 4.11).

In 1836, he moved to the parish in Axminster, where the family owned the living. It was while he was there that the great Bindon landslip occurred between Lyme Regis and Seaton on 25 December 1839. This was a gigantic landslide, and it was reported that the mass that moved was 150 by 500 metres and up to 70 metres thick. There was particular interest in a feature that became known as Goat Island—a large displaced block—and a large deep tension crack, which became known as The Great Chasm, at its landward side. This presented him with the opportunity, along with the Bucklands who were visiting him, and Lieutenant Colonel William Dawson, to write a magnificent account of this great catastrophe (Conybeare et al. 1840). They speculated on the mechanism of the landslide, and subsequent

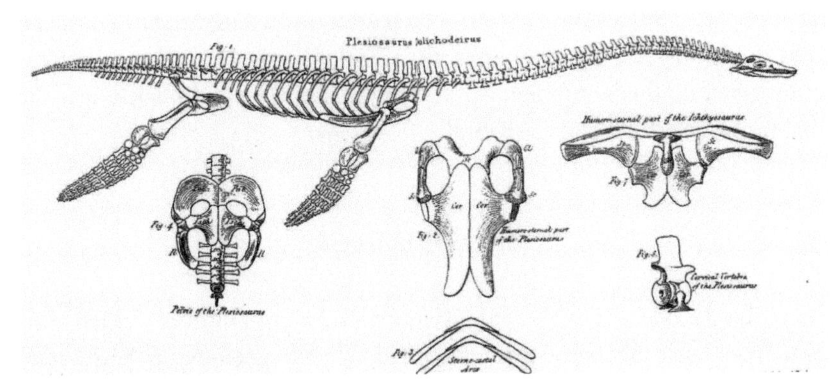

Figure 4.11 Illustration of plesiosaur skeletal anatomy from Conybeare's 1824 paper describing the skeleton found by Mary Anning.
Source: https://en.wikipedia.org/wiki/William_Conybeare_(geologist).

Figure 4.12 Mary Buckland's depiction of the Bindon Landslip of 1839.
Source: https://www.lymeregismuseum.co.uk/collection/the-bindon-landslip-of-1839/.

research has indicated that they were highly perceptive (Gallois 2010). Indeed, this is commonly thought to be the first detailed scientific description of a landslide (Figure 4.12).

In 1844, Conybeare was a recipient of the Geological Society's Wollaston Medal, and then in 1845 he was instituted to the deanery of Landaff, Wales. In July 1857, he was taken ill on the way to Weybridge, Surrey, to see his mortally ill eldest son, William John Conybeare. His own death, from 'pulmonary apoplexy', followed very shortly thereafter on 12 August 1857, at Itchen Stoke in Hampshire, where another son, Charles Ranken Conybeare, had recently taken up the incumbency of the parish church. He is buried near the Chapter House at Landaff Cathedral.

Henry De la Beche

Henry De la Beche (1796–1855) (Figure 4.13), pronounced 'beach' and sometimes spelt 'Bêche', was a major figure in the geology of the Jurassic Coast, not least through his contacts with other geologists, including William Buckland, William Conybeare, and Mary Anning. His father, born Thomas

Figure 4.13 Henry De la Beche with a geological hammer, checked waistcoat, and gold-rimmed spectacles.
Source: https://en.wikipedia.org/wiki/Henry_De_la_Beche.

Beach (1755–1801), had the family name changed to De la Beche in 1790, on the strength of a tradition that they were descended from an ancient family of that name from Aldworth in Berkshire. De la Beche was born in London, though his father, a British Army officer, was a slaveowner with an estate, Halse Hall, in Jamaica. He died while his son was only five years old. Henry's interest in fossils began at school in Keynsham.

De la Beche spent his early life living with his mother in Lyme Regis, where he further developed his love for geology through, *inter alia*, his friendship with Mary Anning. However, as a young teenager he entered the Royal Military College, then at Great Marlow, with the intention of following in his father's military footsteps. However, this was not to be. At the age of 15 he was thrown out of the college for insubordination, though probably not before he had learnt some skills, such as topographic sketching. A change of course was required. After drifting around and living off the income from the Jamaican estate, which he had inherited following the death of his father, he settled in Lyme Regis in 1812, and became acquainted with the Anning family. He became yet more fascinated by the area's geology. In 1817, at the age of 21, he joined the Geological Society of London and was made a Fellow of the Royal Society two years later. He became an avid fossil collector and illustrator (Figure 4.14) and collaborated with Conybeare on an important paper on the anatomy of the ichthyosaur and plesiosaur that was presented before the Society in 1821. In 1818, he entered what was to become an unhappy marriage to Letitia Whyte. The marriage was not a success. He went on a solo trip to Jamaica to attend to estate business in November 1823, and, on his return, discovered that Letitia had left him for another man. A divorce was granted in 1826.

The income from his estates in Jamaica came to an end in 1831, after an uprising of slaves. The British Government decided to abolish slavery in the British West Indies in 1833. De la Beche then ceased to be a wealthy gentleman and began, by necessity, working for a living. He remained a colourful figure, with a fondness for bright-checked waistcoats and gold-rimmed spectacles, and, some say, with a penchant for young ladies. Although he was a slaveowner, he held anti-slavery views, but was more for amelioration than abolition. He favoured gradual elimination of slavery and attempted to ameliorate his slaves' lives by providing them with allotments and granting the right to sell their produce; he also abolished the use of the whip on his estate, and he provided a missionary.

De la Beche's geological studies included many that related to the Jurassic Coast. His first paper, 'Remarks on the Geology of the South Coast of England, from Bridport Harbour, Dorset, to Babbacombe Bay, Devon' (1822), was read before the Geological Society on 5 March 1819. With Conybeare and Buckland he became involved with publications on the fossil reptiles that had been found in the Lias beds around Lyme. In 1826, he published a detailed stratigraphy of the Lias and of the Greensand and the Chalk at Lyme Regis and Beer, respectively. He also produced a geological map of the area (De La Beche, H. 1826: A tourist guide to the geology of Lyme Regis at a scale of roughly 1 inch to the mile [1:63,360]).

His celebrated lithograph *Duria Antiquior* (Figure 4.4) (1830), sold in aid of Mary Anning, was, as we have already noted, the first time that the extraordinary discoveries of the new science of geology had been recast into an actual scene of the flora and fauna of a 'lost world'. It shows prehistoric

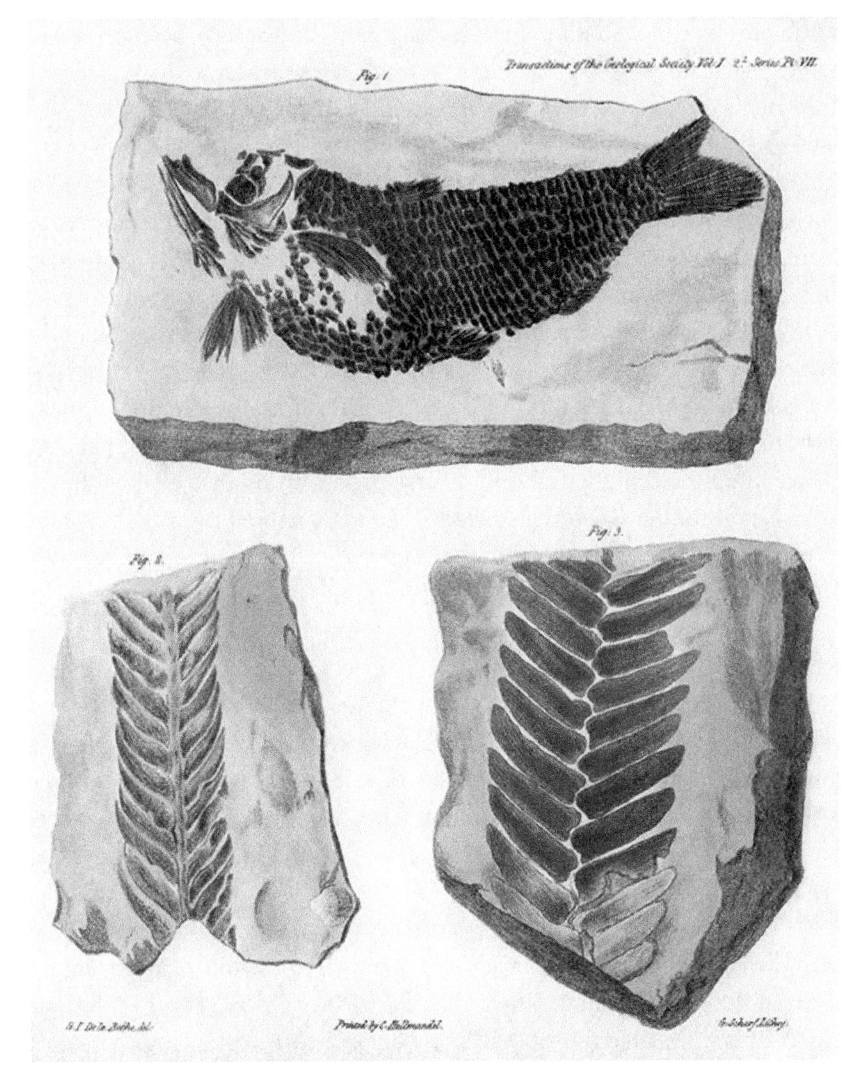

Figure 4.14 Fossils from the Lias of Lyme Regis and Axminster, illustrating De la Beche's finds in the *Transactions of the Geological Society*, 1822. De la Beche was an accomplished artist.

creatures—the plesiosaur, the pterodactyl, and the ichthyosaur—in the air, below the sea, and on land, eating each other and defecating.

With Buckland, he also wrote on the geology of Weymouth and its neighbourhood and was much fascinated by the valleys that furrowed the Jurassic coastline. The Char Valley was enormous yet only contained a small stream at the present time. To him, this suggested that it must have been carved by some greater force—the great diluvial Flood. With respect to the Weymouth area, Buckland and De la Beche discussed the origin of the sarsen stones in the Valley of Stones near Portesham, the hollows created by solution of the Chalk, the fossils of the great dirt bed of the Purbeck rocks on the Isle of Portland, the forces that had elevated and depressed the land, the development of major faults, and the huge amount of strata that had been removed in an area where stream action was perceived to be slight. He also wrote a note in 1826 on a fossiliferous submerged forest bed found near the mouth of the River Char.

De la Beche was a prolific author and wrote about the geology of many parts of the world, including the south of Switzerland, France, and Jamaica. He also published numerous and diverse papers on English geology in the *Transactions of the Geological Society of London*, as well as in the *Memoirs of the Geological Survey*, notably the 'Report on the Geology of Cornwall, Devon and West Somerset' (1839).

Although he gained financial benefit from the institution of slavery when young, De la Beche showed great respect for the less advantaged later in life. He became a staunch advocate of educating all interested citizens, even those of the lower classes. Thus, he wrote various textbooks. He also founded the Museum of Practical Geology, lobbied for the establishment of England's School of Mines, and launched a series of scientific lectures for 'working men', a tradition that was continued by Thomas Huxley (1825–1895).

De la Beche secured financial assistance from the Board of Ordnance to map the geology of Devon at a scale of one inch to the mile. His geological maps of the West Country were published by them in 1834 and 1835 (Bate 2010). This stimulated the establishment of the Ordnance Geological Survey in 1835, later to become the Geological Survey of Great Britain and Ireland in 1845. De la Beche was its first Director. His great success, both in launching and sustaining the Survey, demanded diplomacy, determination, and deviousness. His career marked the beginning of a transition in geology as more geologists began to pursue the science as paid professionals. Having been a gentleman of leisure, he was in a good position to bridge the gap between those who had pursued geological studies in their leisure time and those who earned a living from it. He was knighted in 1848. De la Beche's final years were difficult. Progressively debilitated by paralysis, he spent his last two years confined to a wheelchair. He died in 1855, aged 59 years, and is buried in Kensal Green Cemetery in London. His fellow geological mapper, George Greenhough, was buried there in the same year. Many of De La Beche's archives are lodged with the National Museum of Wales.

5
The Geological Map

At the beginning of the nineteenth century, interest and expertise arose in the field of geological mapping, partly because relatively accurate base maps of topography were being produced by bodies like the Ordnance Survey. In France, the work of Georges Cuvier (1769–1832) and Alexandre Brongniart (1770–1847) on the Paris Basin was especially influential, while in Britain it was the work of William Smith that was crucial (see Smith 2021).

William 'Strata' Smith (1769–1839) (Figures 5.1 and 5.2), born in Churchill in Oxfordshire and son of a farmer, received a scanty elementary education at the village school but trained himself in land surveying and civil engineering. He was a self-taught genius of rare originality and with exceptionally keen powers of observation. He was the first Englishman to recognise fossils in their full significance as a means of determining the relative age of strata. He was not simply content with the determination of a chronological succession of strata; he traced their surface outcrops and thus built up the material for his maps and sections. In 1815, his famous map of England and Wales (Figures 5.3 and 5.4) appeared, consisting of 15 sheets at a scale of 1 inch to 5 miles (*c.* 1:300,000).

As von Zittel (1901, p. 111) remarked: 'Smith's map is the first attempt to represent on a large scale the geological relations of any extensive tract of ground in Europe. It was a magnificent achievement, and was the model of all subsequent geological maps.' He also introduced a stratigraphical terminology based on local names in practical use, including Lias, Forest Marble, Cornbrash, Coral Rag, and Portland Rock. Smith did not do much detailed geological surveying on the Jurassic Coast, although in 1812 he advised the Weymouth Corporation on draining the Backwater at Weymouth.

George Greenough (1778–1855), founder of the Geological Society of London, produced a geological map of England and Wales in 1819, soon after the appearance of Smith's and based in part on it. In Scotland, John MacCulloch (1773–1835) completed a geological map of that country in 1834. It was published posthumously in 1836. Geological map-making became a major activity. This led to the Geological Society negotiating with the

Geological Pioneers of the Jurassic Coast. Andrew S. Goudie and Denys Brunsden, Oxford University Press.
 DOI: 10.1093/oso/9780197638088.003.0005

Figure 5.1 William Smith's portrait by Hugues Fourau.

Source: https://en.wikipedia.org/wiki/William_Smith_(geologist)#/media/File:William_Smith_(geologist).jpg.

government for the establishment of a geological survey. In his Presidential Address in 1836, Charles Lyell was able to announce that the negotiations were successful, that a new geological survey was to be established, and that Henry De la Beche was to be its first head. It was initially to be under the Ordnance Survey. De la Beche was an ideal person because he already had personal experience of producing maps—this included an impressive geological map of Devon which had been produced in 1834 (see Bate 2010). His *Report on the Geology of Cornwall, Devon and West Somerset*, published in 1839, effectively became the first British geological memoir. Over the next two decades, under De la Beche's leadership, the Geological Survey became securely established and appointed a whole series of field geologists.

A number of memoirs were published by the Geological Survey. Their production brought many distinguished geologists to the Jurassic Coast, and these have increased our knowledge of the area's geology immeasurably.

Figure 5.2 The William Smith Medal of the Geological Society of London. It is awarded for excellence in contributions to applied and economic aspects of geoscience.

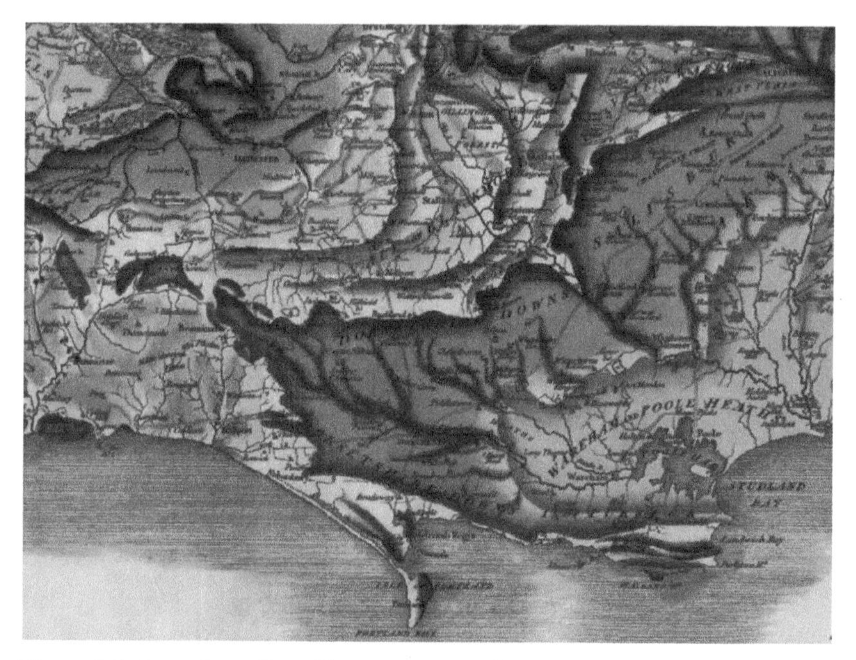

Figure 5.3 A portion of Smith's map showing the Dorset coast.

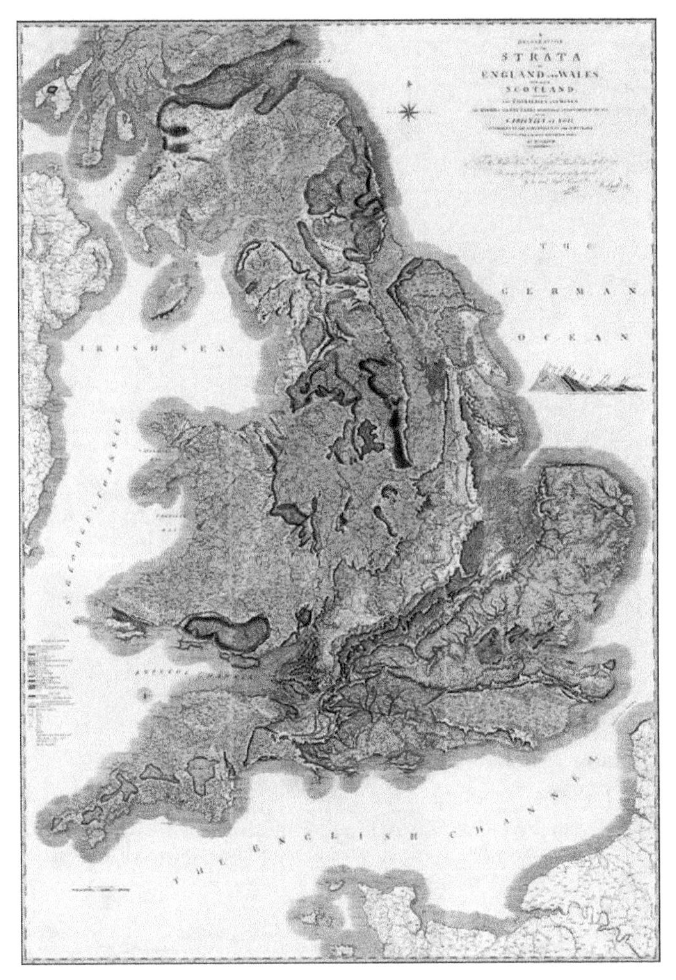

Figure 5.4 William Smith's geology map of England and Wales (1815) placed the geology of the Jurassic Coast in its national context.

One forgotten contributor was Francis Brian Auburn Welch (1903–1987) (Figure 5.5). He was educated at Cheltenham College (locally called the Gentleman's College to differentiate it from the academically superior Ladies' College in the same town) and Bristol University. He joined the Survey in 1931. Between 1931 and 1936, he undertook mapping, and his map of the Bridport area (Sheet 327) is widely admired. He was one of the authors of the memoir on *The Geology of the Country around Bridport and Yeovil*. See also Box 5.1.

Figure 5.5 G. C. Dines, F. B. A. Welch (centre), and G. Kellaway relaxing in the 1950s.

Source: http://geoscenic.bgs.ac.uk/asset-bank/action/viewAsset?id=7398 (BGS Image P008684). We are grateful to the British Geological Survey for permission to reproduce this image.

Box 5.1 Important Memoirs of the Jurassic Coast

Among the most important memoirs relating to the Jurassic Coast are the following:

Arkell, W. J. 1947. *The Geology of the Country around Weymouth, Swanage, Corfe & Lulworth.*

Reid, C. 1899. *The Geology of the Country around Dorchester.*

Strahan, A. 1898. *The Geology of the Isle of Purbeck and Weymouth.*

Wilson, V., Welch, F. B. A., Robbie, J. A., and Green, G. W. 1958. *Geology of the country around Bridport and Yeovil.*

Woodward, H. B., Ussher, W. A. E., and Jukes-Browne, A. J. 1911. *The Geology of the Country near Sidmouth and Lyme Regis.*

Thomas Webster

Thomas Webster (1773–1844) was born in the Orkneys, well away from the Jurassic Coast. He studied architecture in London and was engaged in the construction of the Royal Institution's headquarters in Albermarle Street. In due course, however, he turned his attention to geology.

Between 1811 and 1813, on a geological commission in the Isle of Wight and Dorset for the antiquary Sir Henry Charles Englefield (1752–1822), Webster, using his drafting capabilities, made the first geological map of the region (Figure 5.6) and elucidated the Mesozoic-Tertiary stratigraphy and structural geology of the Hampshire and London basins. In 1814, Webster published a paper in which the fossils in the main divisions of the Chalk were for the first time indicated, and those of the Tertiary strata were shown to prove alternations of marine and freshwater conditions (Woodward 1907, p. 49).

Webster's illustrated descriptive letters in Englefield's *Picturesque Beauties of the Isle of Wight* (Englefield et al. 1816) were highly regarded and helped to establish his geological career. His illustrations are impressive (Figures 5.7 and 5.8). Of particular interest is his drawing of the remarkable fault in the Chalk cliffs of Purbeck between Ballard Point and the Foreland. Subsequently he wrote on the strata between Swanage and Portland. The use of fossils to delimit the stratigraphic succession was highly original and was stimulated by the work done in the Paris Basin by Georges Cuvier and Alexandre

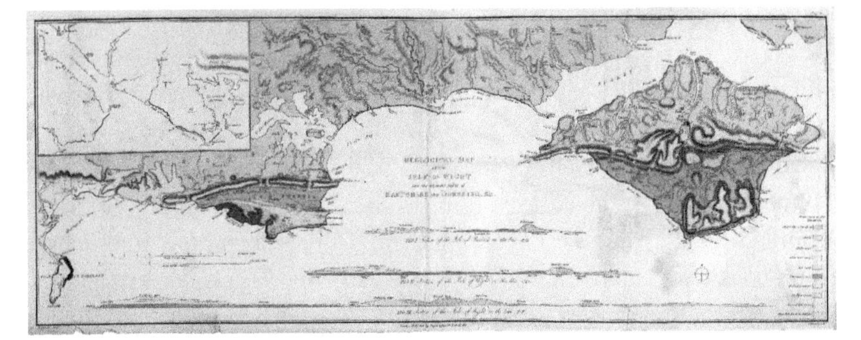

Figure 5.6 Thomas Webster's map of the geology of the Isle of Wight and neighbouring stretches of the Jurassic Coast.
Source: Englefield (1816).

Figure 5.7 One of Webster's drawings of Lulworth Cove, showing the folding of the strata.
Source: Englefield (1816).

Figure 5.8 Webster's drawing of Durlston Head, showing the deformed nature of the strata.
Source: Englefield (1816).

Brongniart. Indeed, he compared his observations in southern England with those that had been undertaken in France.

We owe to Webster many of the geological names we use today. In 1816, he adopted or introduced the stratigraphical names 'Purbeck', 'Portland', 'Kimmeridge Clay', 'Lower Greensand', and 'Upper Greensand' into the geological lexicon (Challinor 1970, pp. 210–213). He was also one of the first people to produce detailed geological maps.

Webster was an early member (1809), later Fellow, of the Geological Society. He worked for the Society in various roles between 1812 and 1827. From July 1827, after failing to gain improved remuneration and terms of employment, he lived by public lecturing on geology, consultancy work, geological illustration, and commissioned writing. From 1841, Webster was the first Professor of Geology at University College, London. The income from his post was minimal, and, by then in poor health, he existed mainly on the charity of geological colleagues and an annual state pension of £50 for services to geology. He died of bronchitis, 'apparently in straitened circumstances', on 26 December 1844, at his lodgings in St Marylebone; he was buried at Highgate Cemetery on 2 January 1845.

Edward Forbes

Edward Forbes (1815–1854) (Figure 5.9), who pronounced his name 'fourbees', was the son of a banker and a native of the Isle of Man. He was

Figure 5.9 The congenial, long-haired, and gregarious Edward Forbes.

Source: https://www.nationalgalleries.org/art-and-artists/43856/edward-forbes-1815-1854-naturalist.

a sickly, long-haired child, but while still young he developed a passion for the study of marine invertebrates. At the age of seven he began collecting specimens for his own natural history museum, and later his father even added a room to their house for them. As Thomas Huxley (1854, p. 1016) related: 'Next, though in very early life, came the perusal of Buckland's "Reliquiæ Diluvianæ", Parkinson's "Organic Remains" and Conybeare's "Geology of England"—rather hard reading that last for a boy, and probably rather wrestled with than understood. These books, however, when he was not more than twelve years old, inspired him with a warm and abiding love of geology.'

Forbes attended the University of Edinburgh, reading both arts and medicine, but failed to take a degree. He became a well-travelled marine biologist noted for his use of dredging of the sea bed. Between 1843 and 1854, he held the posts of Professor of Botany at King's College, London and Palaeontologist to the Geological Survey. It was in this latter capacity that he helped to work out the stratigraphy of the Purbeck Beds of Dorset (1850). In 1845, he was elected a Fellow of the Royal Society, and, in 1853, he was chosen as President of the Geological Society. On 31 August 1848, he married Emily Marianne Ashworth. He was besotted with her and she accompanied him into the field, even after their children were born. She died in 1909.

Although known as a phenomenally hard worker, Forbes was regarded as a gregarious and congenial individual, taking great pleasure in organising and composing light satirical poetry for scientific social events. Forbes used his artistic abilities to create humorous drawings for his publications and was also a frequent contributor to *Punch*. He was also the founder member of three societies: the Maga Club, a student club dedicated to literature and good fellowship; the Universal Brotherhood of Friends of the Truth, whose watchword was 'wine, love, learning'; and the Red Lions, a dining club for younger members of the British Association, named after the tavern where the first meeting was held (Gardiner 1993). He also served as an important mentor to the young biologist Thomas Huxley.

Forbes died of kidney failure aged only 39 years, having just been appointed Regius Professor of Natural History at Edinburgh. At the time of his death he was preparing a monograph on the invertebrate fauna of the Purbeck Beds. Details of his painful death and of the nursing provided by his devoted wife are given in a full-length biography written by George Wilson and Archibald Geikie in 1861. He has been variously credited as being one of the 'fathers' of an impressive range of disciplines, including oceanography, marine biology, palaeoecology, invertebrate palaeontology, and biogeography. Unfortunately, in the minds of some people he is best known for his 'azoic theory', which postulated that marine life did not exist on sea beds at depths greater than 300 fathoms. This was soon to be disproved, but does not diminish the many significant accomplishments of his remarkable but brief career.

The Professor of Botany at Edinburgh, J. H. Balfour (1858, p. 35) commented on his colleague:

> We have lost an original thinker, a careful observer, a correct reasoner, an able writer, a pleasing and painstaking instructor, and a valued friend. His sun is gone down ere it is yet day, and the extinction of such a luminary has cast a shade over the scientific horizon. Truly God's ways are not as our ways, nor his thoughts as our thoughts. Let us learn the lesson which the solemn event teaches, and so number our days as to apply our hearts to heavenly wisdom.

Huxley (1854, p. 1016) also wrote a fulsome obituary of his mentor, remarking that 'Edward Forbes had a great intellect. He was an acute and subtle thinker, and the broad philosophical tone and comprehensive grasp of the many-sided mind enabled him to appreciate and to understand the labours of others in fields of inquiry far different from his own.'

Huxley also remarked (p. 1018) that 'Worked to death, his time and his knowledge were at the disposal of all comers; and though his published works have been comparatively few, his ideas have been as the grain of mustard seed in the parable—they have grown into trees and brought forth fruit an hundred fold; but he never seemed to think it worthwhile to claim his share.' He added, 'The old mourn him as a son, and the young as a brother.'

Henry William Bristow

Henry Bristow (1817–1889) (Figure 5.10), whose father was a Major-General who had served in the Peninsular War, was educated at King's College, London, where he studied civil engineering and applied science. When just 25 years of age, he joined the Geological Survey under Henry De la Beche. He remained with the organisation for 46 years.

Figure 5.10 Henry William Bristow. Luxuriant beards had now become common in the community of geologists.

Source: https://prints.royalsociety.org/collections/geology-crystal-and-mineral-prints-drawings?page=2.

Early in his career he had mapped parts of Wales and the Cotswolds, but then he moved south. He was transferred to Wincanton in 1845, and worked his way through Somerset and Dorset, reaching Swanage on the coast in 1851. Although he published relatively little, the importance of Bristow's work was appreciated by fellow geological surveyor, Horace Woodward (1889, p. 382), who wrote a fulsome obituary.

> [A]nd working eastwards to Lyme Regis, he personally surveyed the greater portions of Dorsetshire.... In the course of this extensive survey all the subdivisions of the Jurassic, Cretaceous and Lower Tertiary strata came under notice; and students who have subsequently paid attention to the structure of these tracts, whether along the fine cliff-sections of the Dorsetshire coast or inland over the Isle of Purbeck, the Ridgway, or Bridport, have borne testimony to the care and accuracy with which Mr. Bristow has depicted the geology. . . . In fact, no one, without actual experience of the process of geological mapping, can fully realize the amount of physical toil and mental labour involved in tracing the geological boundaries and faults in a region where so many subdivisions occur, and where they appear often in irregular and unexpected juxtaposition.

Woodward also noted that Bristow developed the use of geological cross-sections.

> The preparation of the Survey maps, however, was supplemented by numerous sections, longitudinal and vertical, which Mr. Bristow constructed with much skill and neatness to illustrate and explain the geology of the regions he had surveyed. The Purbeck Beds were especially illustrated in this way, and while the palaeontology of these beds was studied by Edward Forbes, the strata themselves were measured in great detail by Mr. Bristow, partly in conjunction with the Rev. Osmond Fisher, and partly with the aid of Mr. W. Whitaker.

Bristow's work was greatly used by Robert Damon.

Bristow and his colleague William Whitaker (1869) described the formation of the Chesil Beach and the Fleet, but their theory for their formation was not very satisfactory and they came up with an implausible explanation involving river erosion between the land the the beach. It was roundly demolished by Codrington (1870).

Bristow was elected a Fellow of the Royal Society in 1862. He retired from the Geological Survey in July 1888, but a sudden stroke led to his death at the age of 72 in June 1889. An obituary in *Nature* mentioned that he was a courteous and gentle man. He suffered from deafness all his life, and this restricted communication and interaction with fellow scientists.

Alfred Jukes-Browne

Alfred John Jukes-Browne (1851–1914) (Figure 5.11) was a palaeontologist and stratigrapher who assisted with work by the Geological Survey along the Jurassic Coast. He was born near Wolverhampton. His uncle was the geologist Joseph Beete Jukes, who had gained a reputation for his work on the English and Irish geological surveys.

He graduated with a BA in Natural Sciences at St John's College, Cambridge. In 1874, he joined the staff of the Geological Survey and was chiefly occupied in mapping parts of Suffolk, Cambridge, Rutland, and Lincoln up to 1883, and then entrusted with the preparation of a monograph on the British Upper Cretaceous rocks. He subsequently wrote a number of books on the subject. He retired from the Geological Survey in 1902, on account of ill health, but continued to write. He was elected a Fellow of the Royal Society in 1909. He died in Torquay, Devon, in 1914.

Figure 5.11 Alfred John Jukes-Browne.
Source: BGS Image no. P585006. Courtesy of the British Geological Survey.

As his obituary in *Nature* (JWJ 1914, p. 667) mentioned,

> Seldom has the triumph of force of will over the most serious disabilities been more strikingly illustrated than in the case of the subject of this notice. To most geologists engaged in field-work the loss of the full use of the limbs would seem to be fatal, but Jukes-Browne, in spite of all difficulties, continued his work as a geological surveyor for twenty years after the almost complete loss of his powers of locomotion.

Also tragically, he had married Emma Jessie Smith in 1881, who died, aged only 34 years, giving birth to their second child in 1892.

He was a contributor, along with Horace Woodward and William Ussher, to the Geological Survey memoir on Sidmouth and Lyme Regis. He was also a noted authority on the Cretaceous Rocks of Britain. In addition to the memoir, Jukes-Browne wrote various other papers relating to the Jurassic Coast. These included a paper on a deep boring made at Lyme (1902). This was of significance, as he reported (p. 279):

> The results of the boring have a twofold interest: first, to all people in the South-west of England they should be a warning against the folly of expecting to find Coal-Measures at a depth of less than 2000 feet below any part of Western Dorset; secondly, they are of interest to geologists for the information which is thus afforded to the great thickness of the Keuper Marls in Devon and Dorset.

He also reported on the geomorphology of the Vale of Marshwood, concentrating on the development of the drainage system and erosion surfaces (1898). Finally, in 1908, he recorded the development of a burning cliff at Lyme (Figure 5.12). Though the locals thought that it might be a volcanic eruption, it was in reality caused by spontaneous combustion of iron pyrites and bituminous shale.

Jukes-Browne also produced maps of the evolution of the British landscape through time, and these included one on the morphology of the English Channel at times of low Pleistocene sea levels and of the state of the Channel at the time of the formation of the Portland Raised Beach (Figures 5.13 and 5.14).

Figure 5.12 Old postcard of the burning cliff of 1908 at Lyme Regis.
Source: https://www.freshford.com/lyme%20regis%20photos.htm.

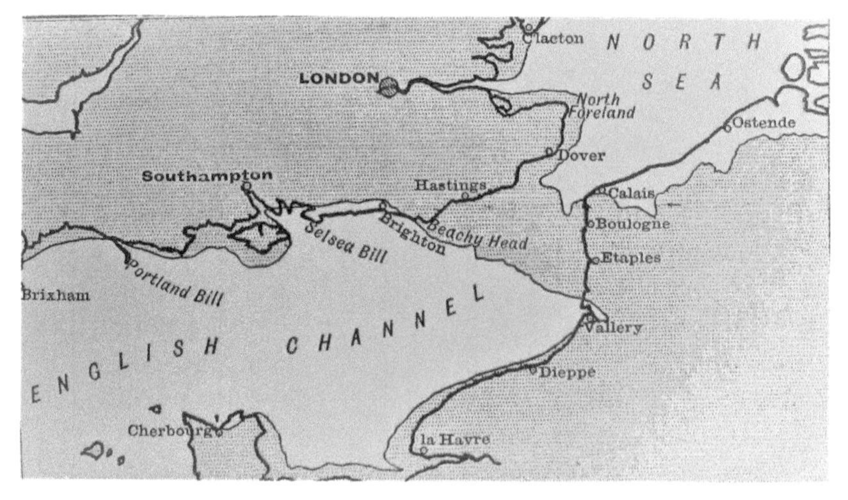

Figure 5.13 Jukes-Browne's map of the English Channel at the time of the Portland Bill Raised Beach.
Source: Jukes-Browne (1892, p. 393).

Figure 5.14 Jukes-Browne's map of the British coastline in the Pleistocene.
Source: Modified from Jukes-Browne (1892, plate XV).

Clement Reid

Older than the deposits of the Pleistocene Ice Ages, but younger than the Cretaceous Chalk are a series of relatively unconsolidated gravels, sands, and clays that are interpreted as being of Paleogene and Neogene age (e.g. dating back to the Eocene, Oligocene, Miocene, and Pliocene). They occur extensively in the Hampshire Basin, but also occur further to the west, as behind

Studland, on Puddletown Heath, at the Hardy Monument in Dorset, and also in parts of Devon, as at Bovey Tracey. One of the prime students of these materials was Clement Reid.

Reid (1853–1916) (Figure 5.15) was a palaeobotanist, born in London. His father was a goldsmith and his great uncle was the chemist Michael Faraday (1791–1867). Reid's love for science is said to have been developed by attendance at lectures for children at the Royal Institution, a body made famous by Faraday's experimental work. His family circumstances meant that he was largely self-taught, but he was nonetheless able to join the Geological Survey of Great Britain as an assistant geologist in 1874, where he was to be employed in drawing up geological maps in various parts of the country.

As a working geologist in the Geological Survey, Reid produced various relevant memoirs on the geology of Hampshire and Dorset between 1899 and 1902, including those on Dorchester, Chichester, Southampton, and Ringwood. He was also notable for his publications on the Pliocene deposits of Britain and, in particular, the Eocene deposits of Dorset. These underlie many of the area's heathlands, and, where they overlie the Chalk, they give rise to large swallow holes, as near Puddletown. He also worked on the palaeobotany of the Purbeck Beds and was an expert on submerged

Figure 5.15 Clement Reid.

Source: J. Flett (1935), The first hundred years of the Geological Survey of Great Britain. Plate VIII (from a photograph by Elliott & Fry, London).

forests along the British coast of the type that occur at the mouth of the Char in Charmouth. He enjoyed linking palaeobotany and archaeology and wrote a pioneering paper on the tufa deposits and kitchen middens of Neolithic age that occurred at Blashenwell, near Corfe Castle. This paper was significant because it demonstrated for the first time the great potential of land snails for dating and identifying past environmental changes. He was particularly concerned with Cenozoic geological deposits and their palaeontology.

He was awarded the Murchison Fund in 1886, won the Bigsby Medal in 1897, and was Vice-President of the Geological Society of London 1913 to 1914. He was elected a Fellow of the Linnean Society in 1888, and of the Royal Society in 1899. He married Eleanor Mary Wynne Edwards (1860–1953) in St Asaph in 1897. She was a distinguished scientist in her own right and worked with Reid on various papers relating to palaeobotany, and she continued on after his death. In 1919, like Clement before her, she was awarded the Murchison Fund. The work of Clement and Eleanor was instrumental in the advance of plant macrofossils analysis as a technique and the important role of biostratigraphy in understanding the Quaternary deposits they encountered. Reid was quiet, retiring, unassuming, and conscientious, and eschewed controversy and debate, though he could be determined in controversy. He retired from the Survey in 1913 and died in Milford-on-Sea, Hampshire in 1916.

Horace Woodward

Horace Bolingbroke Woodward (1848–1914) (Figure 5.16) was the second of two geologists of the Jurassic Coast with that surname—the other being John. He was the son of Dr Samuel Woodward, an expert on mollusca in the department of geology in the British Museum.

Woodward began his geological career in a role at the Geological Society, which he joined in 1867. He worked under Henry Bristow in Somerset, Dorset, and other areas. In 1896, he was elected a Fellow of the Royal Society and, in 1909, he received the Wollaston Medal of the Geological Society of London. He died in Croydon in 1914. An appreciation in the following year (Anon 1915, p. 142) mentioned that his 'life was more generously spent in aiding others than in those original researches which might permanently establish his reputation'. He was said to be 'cheery, tactful, and sympathetic'.

Figure 5.16 Horace Bolingbroke Woodward.
Source: http://earthwise.bgs.ac.uk/index.php/File:P575823.jpg. BGS Photograph number P575823. Courtesy of the British Geological Survey.

Woodward published much important work, including a general text on the Geology of England and Wales (1876); a three-volume work on the Jurassic Rocks of Britain, which appeared in 1892–1895; a geological map of Britain (1904); a magnificent history of the Geological Society to celebrate its centenary in 1907; and a more general history of geology (1911).

His most relevant work with respect to the Jurassic Coast was the Geological Survey memoir on Sidmouth and Lyme Regis, which he wrote with William Ussher and Alfred Jukes-Browne in 1911.

William Ussher

William Augustus Edmond Ussher (1849–1920) (Figure 5.17) was born in County Galway, Ireland. He could trace his roots back to Archbishop James Ussher, who famously stated that the date of creation was 23 October 4004 BC. In 1893, he married Alice Mabel Storrs, a girl 20 years his junior. They had eight children, but this seems not to have stopped his geological activities.

Ussher joined the Geological Survey in 1868, at the age of 19 years, as an assistant geologist under the Director, Sir Roderick Impey Murchison, and he spent most of his career mapping the West Country before retiring in 1909. The Ussher Society, formed in 1962 to act as a focus for geological work in the region, was named after him. The Geological Society awarded him the Wollaston Medal in 1890 and the Murchison Medal in 1914.

Figure 5.17 William Ussher (seated) with co-surveyor and fellow pipe smoker, Horace Woodward, in 1868.

Source: British Geological Survey, P575831. Reproduced with permission of the Geological Society of London.

Ussher carried out most of his work in Cornwall, Devon, and Somerset but contributed to the geology of the Jurassic Coast through his authorship of the *Geological Survey Memoir* on the geology of the country near Sidmouth and Lyme Regis. He also wrote on the sub-divisions of the Triassic Beds in that area (1875, 1876).

His obituary in *Nature* (Anon 1920, p. 144) remarked that 'many British geologists will lose an old friend who, whether in his usual mood of breezy optimism, or in a rarer phase of boisterous pessimism, was always good company'.

Aubrey Strahan

Aubrey Strahan (1852–1928) (Figure 5.18) was born in London and raised at Blackmore Hall, near Sidmouth, until he went to Eton at the age of 13 years. He then progressed to St. John's College, Cambridge in 1870. In May 1875, the year of his graduation, he was employed in a temporary capacity by the Geological Survey, then headed by Andrew Ramsay, as an assistant geologist. He was to remain with the Survey for the rest of his professional life.

Strahan was elected a Fellow of the Royal Society in 1903, and he was President of the Geological Society of London in 1913 and 1914. He was Director of the Geological Survey from 1914 until his retirement in 1920, and he made sure the work of this body was valuable for the prosecution of the First World War. Following Britain's entry into the war on 4 August 1914, the Survey was soon called upon to provide expert geological advice in respect of the British Expeditionary Force's theatre of activity in Belgium and northern France. In addition, advice was sought on obtaining temporary supplies of drinking water at short notice from superficial deposits and from the Upper Cretaceous chalks and Paleogene sands and clays which crop out in the region. They also advised on the appropriate stone to be used for Commonwealth War Graves. Quite properly, they selected Portland Stone.

Strahan lived in Goring-on-Thames in Oxfordshire until his death. During his long career, Strahan contributed to more than 30 memoirs of the Geological Survey, these mostly being detailed descriptions and explanations of the areas covered by individual sheets of the Geological Map. He also published many papers in academic journals. He was created a KBE in 1919, the same year in which he was awarded the Wollaston Medal by the Geological Society of London. His obituary in *Nature* (Anon 1928, p. 461)

Figure 5.18 Aubrey Strahan.
Source: https://en.wikipedia.org/wiki/Aubrey_Strahan#/media/File:PSM_V65_D570_Aubrey_Strahan.png.

mentions that 'he was little disposed to speculation or hypothesis' and that he was 'averse to brilliant and elusive hypotheses'. His Royal Society obituary notice by J. S. Flett noted (1928, p. xx) that 'Painstaking, methodical and diligent, he was much trusted by his colleagues. His admirable administrative abilities soon brought him to the fore, and involved him in much activity that was not strictly scientific.'

Strahan is important because, in 1898, he wrote the geological memoir, *The Geology of the Isle of Purbeck and Weymouth*. He also wrote various papers on the tectonic structures of Purbeck, the nature of chloritic marls at Mupe Bay, and on the Eocene beds found at Bincombe, just inland from Weymouth.

William Joscelyn Arkell

Hugh Torrens, in his 2004 survey of Dorset geologists, stated (p. 12) that 'The Dorset geologist of the 20th century was undoubtedly William Joscelyn Arkell.'

Arkell (1904–1958) (Figure 5.19), a member of the great brewing family, which has brewed in Swindon since 1843, was born at Highworth in Wiltshire, the youngest of seven children. The appropriately named Highworth is the highest town in Wiltshire and located on the Jurassic Corallian ridge. The Arkells were reasonably wealthy and also had a bungalow called 'Faraways' in Ringstead, east of Weymouth, where he spent much of his time.

Arkell went to a preparatory school in Durlston near Swanage, where he revelled in the delights of the neighbourhood. Later he went to Wellington College and then to Oxford University, where he obtained a first-class degree in geology. Following that triumph, he was awarded a Burdett-Coutts scholarship, which enabled him to engage in serious research on the taxonomy of the bivalves from the Upper Jurassic Corallian beds of England. For this and other papers on the Jurassic of southern England, he was awarded a DPhil in 1928.

After appointment as a lecturer at New College in 1929, he was elected a senior research fellow from 1933 to 1940. He soon became a master of the Jurassic. He was still only in his twenties when he wrote in 1933 his great tome (681 pages) on the Jurassic of Britain. He followed this up in 1956 with his massive study of the Jurassic in the world. This was designed as a guide to the Jurassic of particular areas and to each individual stage over the whole world. He synthesised and reviewed critically the information dispersed throughout the enormous literature that existed on this topic; for the first time, a comprehensive global picture began to emerge, forming the framework for future and further elaboration.

Arkell became a principal in the Ministry of War Transport in 1941. In 1943, he contracted a serious chest illness, and for several years he was unable to undertake strenuous activity. In 1947, he was elected a Fellow of the Royal Society and became a senior research fellow at Trinity College, Cambridge.

Figure 5.19 W. J. Arkell. From a painting commissioned by Shell Oil Co. for an International Symposium on the Jurassic System.

Source: https://www.geolsoc.org.uk/Geoscientist/Archive/December_January-2014/In-the-bleak-midwinter. We are grateful to the Geological Society of London for permission to reproduce this image.

With improved health he was once more able to travel abroad, and his studies on the Jurassic ammonites and stratigraphy were intensified.

His major work on the Jurassic Coast was his Geological Survey memoir on *The Geology of the Country around Weymouth, Swanage, Corfe and Lulworth*. This was largely written in 1939, but only appeared in 1947. However, he also

Figure 5.20 Lulworth Cove from the east, showing the upended strata. Arkell was much interested in the tectonics of this area (ASG).

wrote a suite of papers on the area in the main scientific journals, including one on the tectonic processes revealed at Lulworth Cove (Figure 5.20) and at the famous Lulworth Crumple visible in Stair Hole (1938) and another on the structures and mud-slides ('mud-glaciers') at Osmington (1951).

Arkell received the Mary Clark Thompson Gold Medal of the US National Academy of Sciences in 1944, the Lyell Medal of the Geological Society of London in 1949, and the von Buch Medal of the German Geological Society in 1953. His interests were by no means confined to the Jurassic. For four winter seasons, from 1926 to 1930, he accompanied Dr Kenneth Sandford on a survey of palaeolithic archaeology in Egypt. He was also a poet and artist, and his small book of poems, *Seven Poems*, appeared after his death in 1958 from a second stroke.

Arkell often expressed strong views in print and was not averse to controversy. He made fairly strident comments about Sydney Savory Buckman, who had earlier worked on ammonites, and on Sir Aubrey Strahan, who had written the previous version of the Geological Survey memoir on the

Weymouth area. As Leslie Reginald Cox wrote in his *Biographical Memoirs of the Royal Society* (1958, p. 7),

> His pen was fluent and his prose style clear and unembellished. He had firm convictions and was not slow to express his opinions on many subjects in unequivocal language; but it was never his intention to give offence, for he bore no ill-will towards those who disagreed with him. He preferred facts to theories; philosophical speculations entered very little into his writings and had no influence on his conclusions.

In his memoir Arkell made a passionate plea for the conservation of the coastline—something achieved more than four decades later when the Jurassic Coast was declared a World Heritage Site. His plea chimed with that made at the far western end of the Jurassic Coast by Vaughan Cornish at much the same time. Arkell wrote these words (1947, p. 9):

> What will be the future of this 30 miles of coastline, so richly endowed as a training ground and museum of geology? Few tracts of equal size could raise so many claims, scientific, aesthetic and literary, for preservation as a national park. At present, however, it seems that little can be done to save it falling piecemeal before the builder. Weymouth and Swanage are expanding apace and must continue to do so. In the past five years the rural road to Portland Ferry has become a street. Wyke Regis has been engulfed. The villas are marching out to Chickerell and round the back of Lodmoor. They have captured Jordan Hill. The next to fall will be Redcliff. A new building estate with unlimited possibilities has appeared at Ringstead: a red-roofed villa has sprung up on the skyline of the White Nothe itself. If the English of the present generation allow this heritage of the community to be irreparably spoilt for private gain they will be held by posterity to have been unworthy to possess it.

6

Gifted Stratigraphers and Palaeontologists

Such was the splendour of the geology of the Jurassic Coast that, during the nineteenth and twentieth centuries, it attracted a large number of gifted geologists. Some of these people used Lyme Regis and Charmouth as their base and, in particular, made very detailed and significant contributions to our understanding of the Lias. These included Muriel Arber, William Dickson Lang, Leonard Spath, and Edward Day. Mention also must be made of an aristocratic pupil of Buckland and Conybeare, Sir Philip de Malpas Grey Egerton (1806–1881), who worked on the fossil fish from Lyme and over two decades published his results in major journals. For many years he mixed his interest in fish with being an MP for various Cheshire constituencies. An alumnus of Eton and Christ Church, who attended lectures by Buckland and Conybeare, he was elected an FRS in 1831 and received the Wollaston Medal in 1873. His collection of fossil fish is in the British Museum (Bettany 2004).

Sydney Savory Buckman lived for a while near Sherborne in northwest Dorset and made major, but controversial, contributions to our knowledge of Jurassic fossils. Another important figure for a short time was William Branwhite Clarke (1798–1878). He was educated at Cambridge, elected a Fellow of the Geological Society in 1826, became a clergyman in Poole in 1833, and published three major papers in the *Magazine of Natural History* on the Chalk of Ballard Down, on the strata between Durlston Head and Old Harry Rocks, and on those of Studland Bay. His contributions on Dorset were cut short by his voluntary migration to Australia in 1839.

Most recently, one needs to acknowledge the work done by Michael House (1930–2002), who was born in Blandford, died in Weymouth, and produced various editions of the ever-valuable *Guide to the Dorset Coast* and a host of papers.

Geological Pioneers of the Jurassic Coast. Andrew S. Goudie and Denys Brunsden, Oxford University Press.
 DOI: 10.1093/oso/9780197638088.003.0006

William Henry Fitton

William Henry Fitton (1780–1861) (Figure 6.1) was a physician and geologist. Born in Dublin, he was a classical scholar at Trinity College. On one occasion, while collecting fossils near Dublin, he was arrested on suspicion of being a rebel—a Fenian subversive. He was carrying a geological hammer, which was considered by the authorities to be a dangerous weapon.

Initially, like so many other geologists of the time, he intended to enter the church, although, as we have seen, he was already interested in mineralogy—an interest which led to travelling to Cornwall and Wales before 1807. In 1808, Fitton entered Edinburgh University to study medicine; he graduated in September 1810. From Scotland he moved to London and attended his first meeting of the Geological Society late in 1810. The following year he set out to establish himself in medical practice. In 1815, he was elected Fellow

Figure 6.1 William Henry Fitton.

Source: https://commons.wikimedia.org/wiki/File:William_Henry_Fitton.jpg.

of the Royal Society and, in 1816, a member of the Geological Society. On 8 June 1820, Fitton married Maria James, who brought him sufficient wealth to retire from medicine, return to London, and devote his time fully to geology. The couple later had eight children (one less than the Bucklands).

Between 1822 and 1824, the period immediately following publication of William Conybeare's and William Phillips's influential *The Geology of England and Wales* (1822), Fitton proved to be an energetic secretary of the Geological Society. Most importantly, he carefully investigated the stratigraphy of rocks below the Chalk in the south of England, the stratigraphic relations of which were at that time still in a confused and debated state. Fitton started publishing the results of his researches in 1824, and these culminated in his huge and detailed memoir of 1836, *Observations on some of the Strata between the Chalk and the Oxford Oolite, in the Southeast of England*. This clearly defined the confused positions and differing marine to freshwater natures of these strata. He recognised that the Gault separated the Upper and Lower Greensand. He gives full details of the Purbeck and Portland strata of the Jurassic Coast (pp. 208–232), and he introduced the term 'Portland Sand' for the clayey, silty, partly dolomitic material between the Kimmeridge Clay Formation and the Portland Group rocks. His descriptions of the petrified trees found on Portland are especially informative. He illustrates one that is over six metres long. The work was rightly hailed as a model of stratigraphic geology. He was awarded the highest honour of the Geological Society—the Wollaston Medal—in 1852.

Fitton deserves to be known for one more significant act. It was he, as much as anyone, who made sure that due recognition was given by the geological establishment to the genius of William Smith. He died on 13 May 1861, at his home in Sussex Gardens, near Hyde Park in London. Sir Roderick Murchison (1862, xxxiv) wrote a fulsome obituary, remarking that Fitton was 'single-minded, guileless and affectionate', and that he was 'so good a man and a sound geologist'.

Joseph Prestwich

Joseph Prestwich (1812–1896) (Figure 6.2) was born in Clapham, London, the eldest surviving son of Joseph Prestwich, a wine merchant descended from a long line of Lancashire landowners, and his wife, Catherine. He attended University College, London, in 1828, to study chemistry and

Figure 6.2 Sir Joseph Prestwich.
Source: https://en.wikipedia.org/wiki/Joseph_Prestwich#/media/File:PSM_V52_D158_Joseph_Prestwich.jpg.

natural philosophy. In 1830, at the age of 18 years, he joined the family business in the City of London. Geological research was conducted in his spare time. Most notably, in 1859, he went to northern France, where he played a pivotal role in determining the unequivocal antiquity of man. In 1870, he married Grace Anne Milne. She was a geologist in her own right and helped him hugely.

In 1872, at the age of 60 years, Prestwich retired from his business life to devote himself to science. He worked in Dorset in 1873. In 1874, he was offered and, after some hesitation, accepted, the chair of geology at the University of Oxford. He held this chair until 1888, working in the museum, leading field excursions, and giving regular courses of lectures. To go from being a wine merchant to holding a chair in Oxford was a singular triumph. He is remembered in Oxford because of the erection of a blue plaque in Prestwich Place. More recognition was to follow: Prestwich was awarded an Honorary Doctor of Civil Law by Oxford in 1888, and he received a knighthood in 1895. Alas, he was too ill to be able to receive it in person from Queen Victoria. He died the following year.

His main contributions to our understanding of the Jurassic Coast relate to his work on the superficial deposits of the Isle of Portland (1875a,b) (Figure 6.3). He discussed the stratigraphy of the raised beach near Portland Bill. This feature, at 15 to 16 metres above mean sea level, is perhaps the most important exposure of this type on the coastline of southern England. Indeed, Prestwich (1875, p. 43) was of the opinion that 'The Portland raised beach is by far the most interesting one in the south of England, whether for its extent, its thickness, its large exposure, or its general conditions.' Shortly before his death he reviewed the whole extent and history of raised beaches along the southern coast of England and even extended his researches to other parts of western Europe, including the Mediterranean. He concluded that the south of England and much of Europe had been submerged to a depth of not less than about 300 metres between the Glacial (or Post-glacial) and the recent or Neolithic periods. This is not a view supported by subsequent research.

Prestwich also noted the presence of mammaliferous drift containing horse and elephant fossils at the north end of Portland near the Verne. He was intrigued by the source of the drift—some of which was composed of rocks not present on the island—and deduced that it was derived by a stream

Figure 6.3 The raised beach at Portland Bill (ASG).

running from the Greensand and Tertiary strata from the hills between Upwey and Dorchester, before the intervening strata were removed by erosion. He espoused catastrophist views about the origin of these deposits and has sometimes been called the last of the catastrophists. He also contributed to the debate about the origin of Chesil Beach (1875).

Osmond Fisher

Osmond Fisher (1817–1914) (Figure 6.4) was a geologist. He had a long life—he was born two years after the Battle of Waterloo and almost lived to see the start of the First World War.

Fisher was born in Osmington, east of Weymouth. His father, John, who was Vicar of that village, an archdeacon and a friend of John Constable, died when Fisher, the eldest of six children, was only 14 years old. He was looked after by his uncle, the Reverend William Fisher, Rector of Poulshot

Figure 6.4 Osmond Fisher.
Source: Geological Magazine (1900).

in Wiltshire. He encouraged Osmond's scientific and mathematical interests. This included collecting fossils from the local Jurassic Coral Rag. He was sent to King's College, London, to pursue mathematics, but he also attended lectures in geology given by Charles Lyell. In addition he took a degree in mathematics at Jesus College, Cambridge, where he later became a tutor. He was ordained into the Church of England and, in 1845, was appointed Curate-in-Charge of All Saints, Dorchester, where he was involved with the founding of the Dorset Museum. However, most of his 39-year career was spent as Rector of Harlton near Cambridge. In 1867, he was left a widower with five sons to bring up, a few weeks after moving there.

Fisher worked on an array of phenomena in Dorset, including the Purbeck Beds, on which he assisted Henry Bristow; faulting along the Ridgeway between Dorchester and Weymouth; elephant remains at Dewlish (north east of Dorchester); the arrangement of Cenozoic beds at Bincombe; the solution hollows in the Chalk near Puddletown; and the origin of Chesil Beach and the Fleet. He thought that the Fleet was the eastern part of a submerged valley and that its former western side had been encroached upon and destroyed by the waves of West Bay. In this he was wrong: it was never an estuary of this type.

Fisher is, however, now best known for his controversial and revolutionary views on the geophysics of the Earth's crust. He used mathematics in an attempt to provide some answers to this. He countered the deduction of William Thomson (1824–1907), later Lord Kelvin, that Earth is completely solid. Instead, he suggested that the behaviour of the Earth's surface must imply the existence of a plastic substratum below the crust. He argued for the probable existence of convection currents. He suggested that the oceans must expand by the addition of volcanic rocks in mid-ocean and that descending convection currents at the continental margins would cause contraction and the formation of fold-mountains in those regions. Many at the time regarded some of these views as wild speculations, but now we see that he was in effect an exponent of the present fundamental paradigm of the Earth Sciences: plate tectonics. His work in this direction culminated in what is generally regarded as the first textbook of geophysics, *Physics of the Earth's Crust* (1881).

He kept writing geological papers until 1913 and died the following year, aged 96 years. His work was appreciated in his lifetime. He received the Murchison and Wollaston Medals from the Geological Society of London and was elected in 1878 as an Honorary Fellow of King's College, London, and, in 1893, of Jesus College, Cambridge. As Charles Davison, a seismologist, wrote in the *Geological Magazine* of 1900 (p. 54),

> Some time or other, no doubt, Mr Fisher's chief life-work will be weighed in the balance. Whether it is found wanting or not, no one will dispute that he has solved one problem with complete success. However bitterly he may have been attacked, his courtesy has never failed. He is one of the few men whose part in controversy has enriched, and never degraded, science. He can look back on a long life of fruitful labour and of kindly service to his fellow-men. At the same time he can reflect that he has never written a harsh word that he could now wish to be withdrawn.

Edward Day

One of the most important students of the Lias beds of the Jurassic Coast was Edward Cecil Hartsinck Day (1833–1895). He enrolled at London University, and, from 1859 to 1860, he attended the Royal School of Mines in London and took up a geological career. To advance this, he moved to Charmouth in 1861, where he was helped by a local fisherman, Tom Hunter. He was elected a Fellow of the Geological Society in the same year and published an important paper on the Lias of the Dorset Coast in 1863. In this he identified a new shell bed in the Middle Lias, which is now called 'Day's Shell Bed' after him. The ammonite *Dayiceras* was also named after him by Leonard Spath following a suggestion from William Dickson Lang 'to commemorate his important work on the Lias at Charmouth.' In 1864, Day purchased 'one of the most perfect Plesiosauri ever found on the Dorsetshire coast' (https://www.freshford.com/Charmouth%60s%20early%20fossil%20hunters.html) (accessed 3 June 2020) for £40 from the local dealer Samuel Clark of Charmouth (Figure 6.5). The most remarkable feature of this discovery was that Day sold it on to the British Museum for £200, a princely sum at the time. The Plesiosaurus can still be seen in the Natural History Museum in London. In 1865, Day was involved with establishing the geology of a possible railway tunnel under the Straits of Dover. He left Charmouth for America in 1867, and was Assay Master in the newly established Columbia College School of Mines. In 1872, he became Professor of Natural Sciences at the Normal College, New York. He died on 4 January 1895, in Algiers, whilst on a tour in an attempt to recover his health.

Figure 6.5 Day's Plesiosaurus (from a lithograph by Richard Owen in 1865).
Source: https://commons.wikimedia.org/wiki/File:Archaeonectrus.jpg.

Day made a thorough revision of the Lias stratigraphy and a section of it that had been developed by Henry De la Beche some three decades earlier, remarking (1863, p. 279) that

> [w]hen I undertook the investigation of the geology of the coast in the neighbourhood of Lyme Regis, I soon found that that section, though according well with certain general facts, was, owing to the rapid progress of geological knowledge, and the consequent changes in our system of classification, utterly useless to students. The divisions and subdivisions of De la Beche were founded solely on lithological grounds, whereas those of the present day are based in great measure upon palaeontological considerations.

He described sections from Lyme to Burton cliffs, and also discussed the organic remains, such as belemnites and ammonites.

Sydney Savory Buckman

Sydney Savory Buckman (1860–1929) (Figure 6.6) was one of foremost authorities on ammonites, the iconic fossils of the Jurassic Coast. Born in Cirencester, Gloucestershire, he was educated at Sherborne School and his

Figure 6.6 S. S. Buckman in 1906.
Source: S. S. Buckman, Type ammonites (Vol. 3).

father was James Buckman (1814–1884), Professor of Geology, Natural History, and Botany at the Royal Agricultural College.
Buckman's father (Figure 6.7) wrote extensively on the fossils of the rocks in the Cotswolds and neighbourhood and was a friend of Charles Darwin. Resigning from his post after prolonged conflicts with a newly appointed reverend principal, probably because of his overt support for Darwin's recently published theory of evolution, he moved in 1863 to farm 'on scientific principles' at Bradford Abbas, a village near Sherborne in Dorset. By then a recognised authority on matters agricultural, his father became active as a teacher of private agricultural students, helped to found the Dorset Natural History and Antiquarian Field Club, and wrote a multitude of articles, both popular and learned. He died in Bradford Abbas on 23 November 1884.

This parental background and the exposure to the hugely fossiliferous strata of northwest Dorset helped to ensure the career path that Buckman would follow. At the age of 18 years he was already the author of a paper on the *Astartes* (fossil clams) of the Inferior Oolite, which he followed up at the age of 21 years with another paper on the ammonites. These studies, says Arkell (1933, p. 19) 'were only the prelude to activities of body and mind which were to revolutionize much of the accepted geological thought of his time. After making himself master of his own district, the rich and intricate Sherborne country, he set forth to apply his restless mind to ever-expanding

Figure 6.7 A portrait of James Buckman from about 1883, when he was living at Bradford Abbas. The portrait was made by a lady geologist in the neighbourhood and given to Cecil H. Hooper when staying on Easter vacation with Professor Buckman. It was later presented to the Royal Agricultural College (now University) at Cirencester.
Source: https://raulibrary.wordpress.com/2019/01/08/james-buckman-rac-professor-1848-62-his-influence-on-charles-darwin-and-benjamin-vicuna-a-chilean-statesman/.

problems, until nothing less than the whole British Jurassic System, to the far north of Scotland, was his undisputed province.'

Buckman's financial background appears somewhat precarious. In 1882, he began farming in Gloucestershire, a profession he abandoned in March 1886 after being defeated by the great agricultural depression. He continued to live in or near the Cotswolds, but started a new career as a novelist, commercial writer, and dealer in fossils in order to support his scientific work on Jurassic rocks and fossils. He had broad intellectual interests. For instance, in 1910, he published *Mating, Marriage and the Status of Women.* He also wrote articles on such topics as alcoholism, women with beards, pneumatic tyres, babies and monkeys, and the idiocy of feminine dress. He and his wife, Maude, promoted cycling by women and advocated that they should be allowed to wear bloomers, knickerbockers, and other sensible forms of attire.

Hugh Torrens reports that in 1904 Buckman suffered a breakdown in health caused by overexertion as a cycling geologist. He moved to Thame, in Oxfordshire, so that he could be closer to scientists based in London and Oxford. However, the move brought a loss of contact with the field-work which had been so critical to the quality of his early work. In 1909, he commenced publication of *Type Ammonites* (starting initially with those from Yorkshire) as a private venture without interference from referees. The work reached seven volumes by the time of his death.

Buckman was undoubtedly a major figure in the use of ammonites to date sequences of strata. He showed that rocks of the same lithology so dated could be of different ages and that sequences of rocks could be incomplete due to periods of non-deposition or subsequent phases of erosion. This was all very positive. Hugh Torrens wrote that Buckman's scientific work before 1904 was of a quality quite unequalled for its time, and John Callomon (1995, p. 134) has averred that 'Buckman's analysis stands as one of the all-time classical landmarks of stratigraphy.' His reputation was such that he received the Murchison Fund and the Lyell Medal of the Geological Society of London.

On the other hand, Buckman appears to have recognised more specific and generic names for his ammonites than other workers thought were justified. In other words, he was a splitter. This created opposition from the lumpers. He was sharply attacked by Arkell for some of his later work (1933, p. 36).

> That Buckman, who had tramped the Cotswolds and the Sherborne country from end to end and knew every quarry intimately, whose earlier work was built up solely on sound field-work, could also be the author of his last paper on *Some Faunal Horizons in Cornbrash* and of some of the later parts of the fifth and sixth volumes of *Type Ammonites*, is difficult to believe. Without any practical knowledge of the Cornbrash, without describing so much as a single section, he proceeded to divide it up into 11 brachiopod zones and coined for it 5 new stage names. Neither zones not stages have any foundation in fact . . . the definitions of all the stage-names, without exception, are fanciful.

Figure 6.8 Thorncombe Beacon, S. S. Buckman's final resting place (ASG).

Buckman died at his home at Long Crendon, in Buckinghamshire, on 26 February 1929. In 1933, his ashes were spread over Thorncombe Beacon (Figure 6.8), Dorset. His obituary notice by Morley-Davies (1930) refers to his personal virtues: 'simplicity of habits, kindness of heart and a keen sense of humour'.

William Dickson Lang

William Lang (1878–1966) wrote his first paper on the geology of the Jurassic Coast in 1903 and competed his final paper just 60 years later. Nearly 50 of the papers related to Charmouth and its vicinity, including the Vale of Marshwood. He was the foremost expert on the stratigraphy of the layers and ledges of the Lias rocks near Charmouth (Figure 6.9) but also contributed to our knowledge of an array of other local phenomena, including landslips, river history, mudflows, and the submerged forest at the mouth of the River Char. He also wrote historical studies of local figures, including Mary Anning and James Harrison.

Lang was the second son of Edward Tickell Lang and Hebe, daughter of John Venn Prior. Lang was born at Kurnal in the Punjab, where his father was a civil engineer. His father shortly thereafter returned to England but died of meningitis in 1880. William was educated at Harrow and then at Pembroke College, Cambridge, where he took the Natural Sciences Tripos, with zoology as his first subject and geology only a secondary one. In 1898, he made his

Figure 6.9 The coast between Lyme and Charmouth showing the Lias ledges and layers (ASG).

first visit to Charmouth and developed a friendship with a local girl, who ten years later was to become his wife. He visited Charmouth each year, partly for geology and partly for romance. Lang gained an appointment at the Natural History Museum in London in 1902 as a geologist. He remained there until his retirement in 1938.

Lang was a very shy man, and this was something of which he was conscious. This created problems for him as an administrator, but he was recognised as being kind, humorous, and possessing integrity. When he retired, aged 60 years, he immediately moved to his house, 'Lias Lea', in Charmouth and may never again have left the West Country. He became an institution in the town and a pillar of the local church. He was President of the Dorset Natural History and Archaeological Society from 1938 to 1940, and a member of its council from 1956 to 1966. He gave much assistance to other geologists, and this is frequently acknowledged, by, *inter alia*, Muriel Arber.

Various other geologists had written about the Lias before Lang. Examples include Henry De la Beche, Edward Day, and Horace Woodward. Lang, however, mapped the strata and established their sequence in extraordinary and unparalleled detail, mainly by careful analysis of ammonites and other contained fossils. A most important paper, dealing with both Dorset and Devon, appeared in 1924.

In this, one can see the immense amount of data that had been acquired by extensive and long-continued fieldwork over many years. It also contains a map of the local geology at the remarkable scale of 1:2500. Equally important is his 1914 paper on the Charmouth cliffs and beach. This, too, provides the finest detail of the strata and also suggests the presence of a valley bulge, a type of feature often ascribed to deformation of susceptible rocks under periglacial conditions (Figure 6.10).

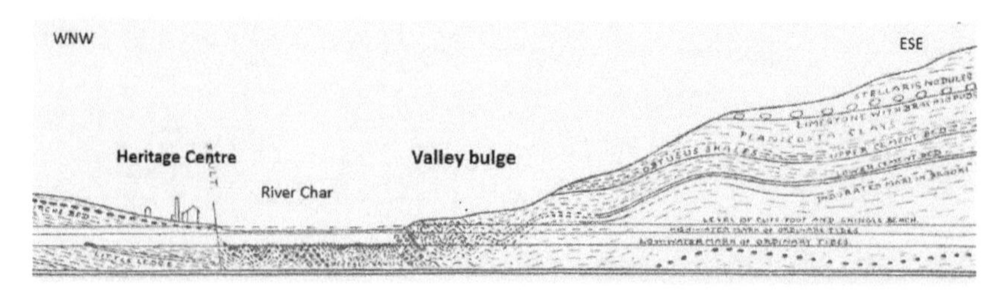

Figure 6.10 A section, modified after Lang (1914), across the mouth of the Char Valley, showing the possible presence of a valley bulge, indicated by contorted strata.

Leonard Spath

Leonard Frank Spath (1882–1957) (Figure 6.11) was often described as an ammonitologist. He made research on ammonites his life's work and studied them, not only in Dorset, but all over the world. He gained a BSc degree in geology at Birkbeck College in 1912, and he obtained employment at the British Museum as an assistant curator in the geology department. In the Great War he served with the British Expeditionary Force in France and Belgium as a private in the Middlesex Regiment. He gained a DSc from the London University in 1921, and he was a lecturer in geology at Birkbeck. He became an FRS in 1940 and won the Lyell Medal of the Geological Society of London in 1945. Spath was born of German parentage in South Africa, but

Figure 6.11 L. F. Spath (from Cox 1957).

Source: https://royalsocietypublishing.org/doi/pdf/10.1098/rsbm.1957.0015. The copyright owner of this image is not clear, but we believe because of its age it is now in the public domain.

Courtesy of the Royal Society.

he kept this secret from both family and colleagues until his death. For him, German ancestry was a sort of stigma (Spath and Wright 1982).

His biographer, Leslie Reginald Cox, noted in *Nature* (1957a, p. 847),

> Spath was so constantly absorbed in his work that it is not surprising that he found little time for other activities, and was rarely, if ever, seen at social functions or even at meetings of learned societies. In his evening capacity of university lecturer he was a painstaking and lucid teacher, much liked by his students. In the Museum, also, he gave unstinted help to younger workers and visitors from overseas.

He wrote voluminously and completed more than 130 papers on ammonites. He was encouraged in his work by William Lang and succeeded Sydney Savory Buckman as the leading British authority on ammonites. In another memoir for the Royal Society, Cox wrote (1957b, p. 221),

> Among Spath's outstanding qualities were a very subtle appreciation of form and a retentive visual memory. Both may have been intensified by his early training as an artist. The first enabled him to appreciate differences in ammonites scarcely definable in words, and the second enabled him, when examining a specimen, quickly to recollect the most closely comparable figure in the literature. No visitor, submitting some fragment or distorted mould of an ammonite to him for his opinion, could fail to be impressed by the keen and critical glance bestowed upon it, the moment's thought followed by a quick reference to some published figure, and the pronouncement of the name and exact geological age of the specimen. Here was a man thoroughly conversant with his subject.

Given his interests, it is not surprising that he wrote a series of major papers on the ammonites from the cliffs of the Charmouth area (Figure 6.12), some of them with Lang.

Muriel Arber

Although Muriel Arber (1913–2004) (Figure 6.13) spent most of her long life in Cambridge, where she was born, she had a deep affinity for Lyme Regis and wrote a series of papers on the landslips and other phenomena of the area.

Figure 6.12 Charmouth Beach with the mouth of the River Char, Stonebarrow, and Golden Cap (ASG).

She came from a highly academic family. Her father was Edward Alexander Newell Arber, a poorly remunerated demonstrator in the Woodwardian Museum (later Sedgwick Museum) in Cambridge from 1899 until his early death in 1918, at the age of 48 years. He was a palaeobotanist but also did notable geomorphological work, most famously in his *The Coast Scenery of North Devon* (1911).

Her mother, Agnes, was a botanist and a Fellow of the Royal Society. It was therefore not perhaps surprising that Muriel Arber had strong academic tendencies and gained her degree in geology at Newnham College. For a period, she undertook research on brachiopods, but later she became a schoolteacher at the King's School in Ely. In the 1970s, she was President of the Geologists' Association and was also awarded the R. H. Worth Prize of the Geological Society of London. In her later years she suffered from blindness; she died in Cambridge, aged 90 years, on 10 May 2004. She is remembered for her humour, modesty, and extraordinary memory. She was a lady of imposing height. She never married.

Muriel took holidays in Lyme Regis with her parents, and she was influenced by William Dickson Lang, a local authority on the Lias. Following in her father's footsteps, she developed a huge interest in cliffs and their

Figure 6.13 Muriel Arber.

Source: http://rudischmid.com/arber/MAA_obit.html. We are most grateful to Rudi and Mena Schmidt for their permission to reproduce this image.

modification by landslipping. She also wrote about brachiopods, the cliffs, glacial deposits, and sea level changes of North Devon (her father's specialty) and about floods and dust storms near Ely.

In 1941, stimulated in part by the centenary of the Christmas 1839 landslip (Figure 6.14), Muriel published a comprehensive review of the various landslips near Axmouth. She also recalled in that paper the way in which the enterprising local inhabitants made money from the landslip (pp. 266–267).

> [A]n increasing number of visitors came to the scene, and by charging an entrance fee to their land the farmers of Dowlands and Bindon were soon more than compensated for the loss of their ground. As many as a thousand tickets were sold in one day, steamers brought parties from Weymouth and Torquay, Mrs. Critchard returned to her ruined house to supply refreshments, and the idea of the landslip so caught the popular fancy that a dance called the Landslip Quadrille was published with a lithograph of the chasm on the title-page. The excitement culminated in August, 1840, when

Figure 6.14 Title page of Ricardo Linter's *Landslip Quadrille* (from Pitts and Brunsden, 1987, fig. 8).

> a grand rustic fete was arranged in connection with the reaping of the corn in the isolated fields 'by attendants of Ceres'. The crowd was so great that few people could see the ceremony and many could not obtain refreshments in spite of the generous supplies of provisions. The 'wagon-loads of delicious food' spread out on the grass, the hams, home-made bread, and water 'boiled in furnaces', are still quoted by Mrs. Gapper, who has herself lived for the last fifty years in the cottage rebuilt near the site of Critchard's, and whose mother was among the farmers' wives who reaped the corn in that harvest of 1840.

She also discussed the controls on landslide occurrence and behaviour (pp. 269–270).

> The distribution of the landslips on the South-East Devon coast depends primarily on the relations of the dips of the beds, and of the unconformity,

> to sea level. Where the plane of junction between the Foxmould sands and the underlying clays occurs in the cliff-section and slopes down in a seaward direction, erosion, by cutting the cliff-face and hence removing the outward support of the beds, will tend to bring about slipping of the upper layers over the lower. This condition is fulfilled in the Hooken Cliff area west of Beer Head, and along the coast from Axmouth to Lyme Regis. At Beer Head itself, and in Whitecliff, between Beer and Seaton, the Triassic-Cretaceous plane of unconformity is almost entirely below sea level. Here, therefore, marine erosion acts on cliffs formed only of Cretaceous material, and the Chalk falls directly to the shore, instead of being carried forward on a sliding base.
>
> As an undercliff develops on the upper surface of the clays, so the factors controlling the slipping become more complicated. Water is impounded, beds are held up by talus, and there are great accumulations of unstable material. The forms of the undercliff and of the inland cliffs which back them are, however, principally dependent on the low degree of the dips, and on the coherence of the upper beds which are let down by the foundering layers of sand. In South-East Devon the chalk and cherty sandstones are massive strata, which may break but do not readily disintegrate. When subsidence occurs, they are therefore capable of sliding bodily, with comparatively little damage, since the dip is sufficiently gentle for the movements not to be shatteringly violent. In this way, solid portions of the cliff face have come to be moved forwards into the undercliff, and the unique character of the landslips of this area is due to the existence of the isolated masses and pinnacles, and above all to the great block of land beyond the chasm at Dowlands.

She also wrote about she wrote about the cliffs and mass movements to the east of Lyme Regis (pp. 280–281).

> The coast-section consists of a continuous sequence of Lower and Middle Lias (clays, shales, marls, limestones and sands), dipping east-south-east, and capped by the overstepping Gault (loams and clays), and the Foxmould (sands) and Chert Beds of the Upper Greensand. Water is held up at many horizons, at each of which slipping is liable to occur. A water-bearing horizon, above which the clays are washed out, occurs near the top of each of the series of the Lower Lias, so that the cliff-face consists of alternating terraces and precipices. This process has caused considerable damage and

> loss of land in the Langmoor Gardens and the Church Cliffs at Lyme Regis, and on Black Ven. On Black Ven and Stonebarrow, water is held up over the upper surface of the Lower Lias and at certain horizons in the Gault and Foxmould above, and thus the Cretaceous and Middle Liassic beds may be washed down over the uppermost terrace of the Jurassic, leaving an inland scarp separated from the seaward cliffs by a tumbled undercliff. Part of the old coast-road over Black Ven has now been carried completely away, and on Stonebarrow large masses of Middle Lias strata occasionally slide bodily forwards into the undercliff known as Fairy Dell. On the low flanks of the cliffs, sands or drift overlying water bearing clays are worn back so as to form a receding escarpment, thus exposing the upper surface of the clays, which become boggy and founder seawards in avalanches and mudflows. Very little slipping has occurred either on Golden Cap, Down Cliff or Doghouse Hill, where the higher beds are all sandy and so almost equally permeable; or where the beds are of uniform lithology east of the big fault in Watton Cliff. Whereas in South-East Devon there is a single plane of weakness which undercuts the massive beds above and allows them to slide forwards periodically in unbroken blocks, in West Dorset the rapid variations in lithology produce a series of water bearing horizons, above each of which the beds may be washed away, the subsidences exhibiting every gradation from a continuous mudflow to an occasional mass of strata moving bodily. By this process, the cliffs have developed their remarkable terraced profile.

In 1973, she returned to this theme and gave an update on the changes that had taken place over the intervening three decades. For example, she discussed the failure that took place at Stonebarrow on 14 May 1942 (Figure 6.15) (p. 125).

> [T]he crest of Stonebarrow broke away, and a slice about 500 m. from east to west and 18 m. from north to south slid down some 15 m. into the undercliff of Fairy Dell beneath. It left a slipscarp of yellow Upper Greensand, far more brilliant than the summit of the cliff had previously been in its slightly vegetated condition. The part which fell carried down with it two radiolocation huts which had been built on the top of the cliff during the Second World War. The displaced slice came to rest tilted backwards towards the main cliff behind the undercliff, and thus demonstrated that the slipping here was due to rotational shear.

Figure 6.15 A radio-location hut that slipped down Stonebarrow on 14 May 1942 (ASG).

She also described the new mudflows of the Black Ven area and the problems that instability had posed in Lyme Regis itself. In 1946, she wrote about the valley systems behind Lyme Regis and noted the similarity between the valley side slopes and those of the coastal cliffs—both, she opined, were largely being moulded by sub-aerial processes.

Muriel Arber's work on the landslips of the Jurassic Coast (Figure 6.16) provides a stable foundation upon which subsequent workers have built.

Michael House

The Geological Society of London obituary for Michael House (Butcher 2002) (Figure 6.17) (1930–2002) begins,

> From his Dorset Jurassic roots, to which he returned in the last years of his life, Michael House relentlessly pursued the marine Devonian rocks and their ammonoid faunas across the face of the Earth. In so doing, he built a formidable reputation as one of the foremost stratigraphical palaeontologists of his generation.

Figure 6.16 The slips and mudflows of Black Ven looking towards Charmouth, Stonebarrow, and Golden Cap (ASG).

Fortunately, he also published extensively on Dorset geology; indeed, his first published paper was on the Red Nodule Beds near Weymouth. He wrote about the county's stratigraphy and also about some of its chalk solutional features (dolines). So important did this work become that *The Official Guide to the Jurassic Coast* is dedicated to him. He was the author of *Geology of the Dorset Coast* for the Geologists' Association Geological Guides Series, Guide No. 22. (1989). This was preceded by the original guide, which covered Poole to Chesil Beach (1958) and complemented the guide (No. 23) by Derek Ager and W. E. Smith (1965) covering the coast of south Devon and Dorset between Branscombe and Burton Bradstock. The two No. 22 guides became the definitive statement on Jurassic Coastal geology for more than 50 years, until John Cope's guide, *Geology of the Dorset Coast* (2016) was published.

Born in Blandford, and with a father who was a master plumber in Portland, House attended school there before undertaking National Service in the army. He studied at Cambridge, taking a first-class degree in geology in 1954. He worked at Durham as a Commonwealth Fund Fellow and at Harvard, Cornell, and the US National Museum. He then moved to Oxford, where he became Dean of St Peter's College (then often called Pot Hall)

Figure 6.17 Left: Michael House in his Hull office. Right: With Richard Edmonds (in blue).

Sources: Michael House: Becker and Kirchgasser (2002), courtesy of the British Geological Survey. Right: Richard Edmonds, courtesy of Ian West, © 2003).

in 1965. In 1967, he was made Chair of the Hull department. In 1986, he moved to Southampton, where he was acknowledged as a master of systematic palaeontology. He also developed ideas relating the Milankovitch hypothesis of orbital forcing and climatic cyclicity to patterns of sedimentation in the Jurassic of Dorset. The current interpretation of the Jurassic strata of the coast owes much to this leap of imagination.

House was much honoured: he received the Daniel Pidgeon Fund in 1957, the Wollaston Fund in 1964, the Murchison Medal of the Geological Society in 1991, the William Bolitho Gold Medal of the Royal Geological Society of Cornwall in 1970, the Neville George Medal of the Geological Society of Glasgow in 1984, and the Sorby Medal of the Yorkshire Geological Society in 1986. This tells the story of a very accomplished man. He was also a founder member of the Ussher Society in 1965.

On the Jurassic Coast, he left a thought-provoking legacy. He had an unrivalled ability to draw coastal sections of great clarity, which he used to

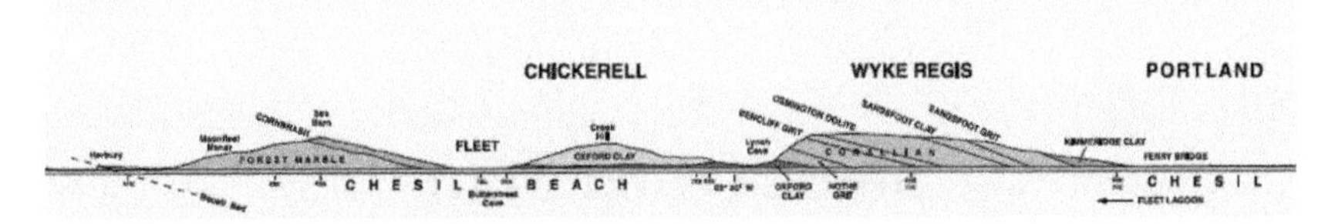

Figure 6.18 An example of a geological cross section of part of the Jurassic Coast drawn by Michael House.

From © Dorset County Council Nomination Document, 2000, p. 60. Courtesy of British Geological Survey, Devon County Council, Dorset Council, NERC (Permit number: IPR/4-2). With permission of the Jurassic Coast Trust.

illustrate the guides to the Dorset coast (Figure 6.18). The sections were finally completed by the World Heritage team for the nomination document for inclusion in the World Heritage listings.

When House was 16 years old, he was much influenced by William Arkell's memoir (1947) on *The Geology of the Country around Weymouth, Swanage, Corfe and Lulworth*. In a letter to Roger Hewitt (http://www.hullgeolsoc.co.uk/hg144.htm), a one-time research student of his, he wrote in 1994,

> You've probably forgotten now that you mentioned Arkell's (Geol. Survey) Memoir published (1947). When I was 16 and ready for something interesting which, since I was living in Weymouth, it certainly provided. What would I be doing now if I had not read it? I sided with Arkell on Kimeridge until I noticed that William Smith used two m's. You are right about how much work still needs to be done in Dorset. Steve Etches has some marvellous KC ammonites with large bites out of the ventral and posterior parts of the body chamber—presumably to eat them better. Have now retired, but doing 12 lectures a year still. I think we will move to Dorset next year.

This quote emphasises the huge enthusiasm that scientists develop for their work and their 'sense of place'. His biography by Norman Butcher (https://www.geolsoc.org.uk/en/About/History/Obituaries%202001%20onwards/Obituaries%202002/Michael%20Robert%20House%201930-2002) concludes with,

> Michael House, who, just like his favourite fossils, enjoyed (in A. Morley-Davies's words) 'a short life and a merry one', was a world figure and a true gentleman.

7
Some Other Stars

Our final group of pioneers of the geology of the Jurassic Coast is a rather diverse one and includes both amateurs and professionals, geologists and non-geologists. They were all characterised by a great range of accomplishments and interests. The role of the amateur was a very significant one, and, as we have seen in earlier parts of this book, a number of the geologists we have described were also clergymen or medics. Some of the geologists we have already discussed were also very respectable artists. So were two of our amateur stars—Peter Orlando Hutchinson and Ernest Burton—while Eleanor Coade made magnificent sculptures.

Eleanor Coade

Eleanor Coade (1733–1821) may seem a strange entry for this book—she was not a geologist. Rather than studying rocks, she made them. She was known as 'Mrs Coade', even though she never married.

She was born and brought up in Devon, although the Coade clan came from Lyme Regis. She started to make the artificial stone that bears her name in London, in 1769. Her factory was located where the Royal Festival Hall now lies. This stone, which has proved to be highly durable, became extremely popular and was used by royalty for monuments in Buckingham Palace, St George's Windsor, and the Brighton Pavilion. Her stone also graces the Jurassic Coast, with examples from Lyme Regis and Weymouth (Figure 7.1). Indeed, a large house in Lyme Regis, named Belmont, was her seaside retreat. The author John Fowles lived there from 1968 to 2005, and it is now owned by the Landmark Trust. It is decorated with the products of her factory in the form of sea creatures, quoins, and series of urns along the parapet; it lies on the corner of Pound Street and Cobb Road.

Coade stone was a special composition of fired clay, carefully mixed with pre-fired raw ball clay carefully mixed with pre-fired, ground-up terracotta or 'grog' and a sprinkling of ground glass, flint, or other silicates. This was

Geological Pioneers of the Jurassic Coast. Andrew S. Goudie and Denys Brunsden, Oxford University Press.
 DOI: 10.1093/oso/9780197638088.003.0007

Figure 7.1 The imposing statue of George III and his lion and unicorn on the Esplanade in Weymouth, erected in 1809, is made of Coade stone (ASG).

shaped in a mould and then passed through a kiln at a very high temperature, making it impervious to rain and frost. Its versatility made it immensely popular for a great variety of sculptures, large and small. It was used by every leading architect and designer of the day, including Robert S. Adam, John Nash, and John Soane. More than 650 surviving Coade stone sculptures have been traced today, not only across the British Isles, but as far afield as Russia, the United States, the Caribbean, South Africa, and Brazil.

Visitors to the coast from London and the Midlands often note the great stags and lions on the gateway arches to the estate at Charborough (to the east of Bere Regis), owned by Richard Grosvenor Plunkett-Ernle-Erle-Drax, MP for South Dorset. These, too, are made of Coade stone. One can also see Coade stone ammonites at the entrance to the Lyme Regis Museum (Figure 7.2). There is a mews in Poundbury named after her.

Peter Orlando Hutchinson

Peter Orlando Hutchinson (1810–1897) (Figure 7.3) settled in Sidmouth at the age of 15 years in 1825. He was a gentleman of leisure who never married,

Figure 7.2 Coade Stone ammonites at the entrance to the Lyme Regis Museum.

Source: https://en.wikipedia.org/wiki/Lyme_Regis_Museum.

Figure 7.3 Peter Orlando Hutchinson, a self-portrait.

Source: https://en.wikipedia.org/wiki/Peter_Orlando_Hutchinson#/media/File:Peter_Orlando_Hutchinson.jpg.

and he became an expert of Sidmouth in all its aspects. He had many talents. His interests ranged from international politics to performing on the flute and French horn in public concerts in Exeter, from carving decorative stonework for the newly restored Sidmouth Parish Church to the latest method of preserving telegraph poles, to name but a few.

He never had any formal geological training. However, he is especially noted for his 1843 book, *The Geology of Sidmouth and of South-eastern Devon*. In this he rambles widely, but makes some extremely valuable observations on the current rates of denudation exemplified by the sediment loads of the River Sid. He believed in the power of present processes rather than in catastrophes to explain the rivers of the area. He also discussed the evolution of the stacks of Ladram Bay. Indeed, he was very interested in coastal change. In addition to his book, he painted and discussed the great 1839 landslip and discussed the various reasons for what some had called the 'Culverhole Catastrophe'. He did not believe it resulted from fire or earthquake, rather that it was caused by water flow. He also wrote about a submerged forest and fossil plants and teeth found at Sidmouth and contributed to a national discussion about coast erosion. Many of his diaries and drawings are available on a web site (https://www.eastdevonaonb.org.uk/our-work/projects/peter-orlando-hutchinson) (accessed 4 September 2021).

John Clavell Mansel-Pleydell

John Clavell Mansel-Pleydell (1817–1902) (Figure 7.4) was born in Bramshaw, Dorset. He entered St John's College, Cambridge, in 1836, and graduated in 1839. He started work as a lawyer, but soon gave up. For 30 years he was an officer in the Queen's Own Dorset Yeomanry. He was promoted from lieutenant to captain on 26 July 1856. He succeeded on his mother's death to the family estate of Whatcombe, Dorset, and to landed property in the Isle of Purbeck in 1863. He helped to found the Dorset Natural History and Antiquarian Field Club in 1875, and was President until his death. In 1876, he was High Sheriff of Dorset and a member of the county council from its establishment in 1887, also until his death. He was also a magistrate. He is noted for his works in natural history: *The Flora of Dorsetshire* (1874; 2nd edn. 1895), *The Birds of Dorsetshire* (1888), and *The Mollusca of Dorsetshire* (1898).

Figure 7.4 John Mansell-Pleydell with terrier by Anthony du Brue.
Source: With kind permission of the Dorset County Museum.

Mansel-Pleydell may be regarded as almost the last of the race of country gentlemen of high social position who took any deep interest in geology. He made geological contributions and presented many geological finds made by himself to the County Museum of Dorset. These included a fore-paddle of a *Pleiosaurus macromerus* and the tusks and molars of a *Mammuthus meridionalis*. Indeed, the Kimmeridge Clay in particular yielded to him many saurian remains, some of which were described by Richard Owen and John Whitaker Hulke (Wills 2013). In 1873, he contributed to the *Geological Magazine*, 'A Brief Memoir on the Geology of Dorset', which provides an excellent overview of the whole geology of the county. He also wrote about the tufa deposit at Blashenwell.

John Coode

Sir John Coode (1816–1892) (Figure 7.5), a Cornishman, was not so much a geologist as a civil engineer. However, he did much to modify the coastline by his major role in the construction of Portland Harbour, the largest man-made feature along the whole coastline. He also wrote in detail about the nature and origin of the Chesil Beach, one of its world-class geomorphological phenomena. He is commemorated by Coode Way, a road in Portland just behind the beach.

In 1847, Coode was appointed resident engineer in charge of the construction of the works at Portland Harbour, which had been designed by James Meadows Rendel. On the death of the latter in 1856, he was appointed engineer-in-chief, a post he retained until the completion of the works in 1872.

Portland Harbour (Figure 7.6) provided the largest area of deep water of any artificial harbour in Great Britain and was a work of major national strategic importance, constructed in part by the use of convict labour. The first stone of the great breakwaters was laid by Albert, the Prince Consort, on 25 July 1849, and the breakwaters were declared complete by HRH Edward the Prince of Wales on 10 August 1872. Coode was knighted in that year for his services in connection with this undertaking. The vast harbour consists of four breakwaters. These have a total length of 4.57 kilometers and enclose approximately 520 hectares of water. He also played a role in the construction of works at Bridport Harbour, West Bay. He went on to build harbours all over the British Empire.

Working on the harbours at West Bay and Portland, which lie at either end of Chesil Beach, it is perhaps not surprising that he took an interest in this huge mass of shingle. His paper on Chesil Beach (1853) is a substantial piece of work and the first detailed discussion on this subject.

Chesil Beach, or Bank, is the most impressive shingle ridge in the British Isles. It is one of five major gravel/shingle features along the British coast,

Figure 7.5 Sir John Coode.
Source: https://en.wikipedia.org/wiki/John_Coode_(engineer)#/media/File:JohnCoode.jpg.

together with Dungeness, Orford Ness, Spey Bay, and Culbin. It is also one of the most impressive shingle accumulations in the world. It is notable not only for its sheer size, but also for its regular crest line, its beautifully even curve, its lack of lateral ridges, and the remarkable—though somewhat complex—grading of pebble sizes that takes place from one end to the other. Starting at West Bay, where the shingle is predominantly pea-sized, to Chiswell on Portland, where the shingle—much of it composed of chert and flint—is predominantly potato-sized, the overall length of the beach is

Figure 7.6 Chesil Beach, about which Coode wrote, and Portland Harbour, which he helped to build.
Source: © Landsat/Copernicus Google Earth, 2016.

28–29 kilometres. Between Abbotsbury Swannery and Small Mouth (where Portland begins), the beach protects a brackish lagoon, the Fleet. At Wyke Regis the beach is more than 200 metres wide, and its height reaches about 15 metres at Chiswell.

Although some antiquarians and topographers, such as Leland and Camden, had noted the splendour of Chesil Beach, Coode's paper was the first scientific, full-length treatment of what he termed (p. 520), 'perhaps the most extraordinary, and at the same time the most extensive, accumulation of shingle, to be met with in this country'. In this paper he described the Bank; provided sections; discussed its geometry, including its parabolic form; gave the results of soundings and borings; traced the source of many of its pebbles to the rocks between Lyme Regis and Sidmouth; stressed the role of waves, rather than tidal currents, in its formation; and discussed the grading of the shingle, and the formation of the 'cans' (gullies) between Portland and Wyke Regis. He mentioned little about the Fleet.

In later years, other scientists turned their attention to Chesil Beach, including Joseph Prestwich (1875), who was critical of Coode's great paper and generated a spirited response from him (Reade et al. 1875, p. 105). Prestwich did not think that the shingle was derived from present-day wave action in Lyme Bay, but believed (p. 78) 'that the shingle of the Chesil Bank is chiefly derived from the materials of the raised beach, of which a remnant still exists in situ on the Bill of Portland, and partly from the harder beds of the Portland and Purbeck formations of that island'. However, Coode's paper was undoubtedly the best researched and most sensible one until new data and techniques became available from the 1960s onwards, as exemplified by the work of Alan Carr and colleagues. Indeed, Carr was to use Coode's surveys to assess rates of recession of Chesil Beach.

Vaughan Cornish

Vaughan Cornish (1862–1948) (Figure 7.7) was born at Debenham, in Suffolk. He was one of those gifted amateurs who made major contributions to national life. He was a geographer of wide interests who studied, *inter alia*,

Figure 7.7 Vaughan Cornish: Geographer and landscape preservationist.

Source: https://upload.wikimedia.org/wikipedia/commons/thumb/2/29/Vaughan_Cornish_LCCN2014716603.jpg/1200px-Vaughan_Cornish_LCCN2014716603.jpg.

waves of water, sand, and snow. In 1898, he produced an important paper on the shingle grading of Chesil Beach. In this he summarised the salient points relating to the increase in shingle size as one moved eastwards from West Bay to Chesilton on Portland (Cornish, 1898, p. 634).

1. The beach is fed at both ends (Bridport and Chesilton).
2. The material fed in at the west end is mostly fine, owing chiefly to the natural groynes at Golden Cap and Thorncombe.
3. The material fed in at the east end is mostly coarse, owing to the nature of the local rock and the mode in which it is supplied to the foreshore.
4. The main drift of water is easterly, but
5. Of the fine shingle carried eastward from Bridport, much is brought back by waves from the east; whereas
6. The strong outset at Chesilton removes such fine stuff as may be there supplied from Portland.
7. The largest waves converge on Chesilton from both sides.

Overall, Cornish was very productive, some might say prolix, producing more than 90 papers and books in a writing career spanning over 60 years. But he was also a pioneer conservationist who did much to help preserve the Jurassic Coast landscape.

Cornish was educated first at home and then, when he was aged 17 years, at St Paul's School in London. After working as a private tutor, he studied chemistry at the Victoria University of Manchester. He gained a BSc with first-class honours in 1888, and proceeded to an MSc in 1892 and a DSc in 1901. In 1891, he married Ellen Agnes Provis (1853–1911). In the same year he became director of technical education for Hampshire. Living in a cliff-top house near Branksome Chine, between Bournemouth and Poole, he began his studies of waves, and, in 1895, he resigned from his job to devote himself to research. In this he was encouraged and financially supported by his wife, who had private means.

Cornish's elder brother, Charles John Cornish, a naturalist and a founder of *Country Life*, was a major influence—in his *Wild England of Today* (1895), he was writing about the threats to clifflands and the pine and heath country of the New Forest. Cornish was also influenced by Sir Francis Younghusband, the Tibetan explorer, who had stirred him in a speech made at the Royal Geographical Society in 1920, in which he stressed the value of studying landscape aesthetics. Thus, he was active in the Council for the

Preservation of Rural England (now called the Campaign to Protect Rural England), founded in 1926, for whom he gave evidence to the National Parks Committee in 1929 and produced *The Scenery of England* (1932). He proposed national parks in *National Parks and the Heritage of Scenery* (1930) and *The Preservation of our Scenery* (1937).

Cornish put his principles into practice after inheriting land in Salcombe Regis, Devon (Figure 7.8), in 1938, described in *The Scenery of Sidmouth* (1940, pp. 73–74). He wrote,

> The full grandeur of the scene from the summit of a cliff depends upon the maintenance of an open space on the landward side to balance the expanse of the sea. Thus not only is a cliff path required, but its users should not be cramped by a high fence or paling on the landward side. . . . Cliff lands are only perfect in their charm when they present a pastoral scene, downland where the shepherd and his flock are wandering, or meadows where the cattle are at graze, or arable where the ploughman and his team are furrowing the field. The most picturesque of all human occupations cannot be seen to advantage where the summit of a cliff is a public playground

Figure 7.8 Vaughan Cornish's idyllic farm in Salcombe Regis.
Source: A photo by Cornish in *The Scenery of England*, CPRE (1932), frontispiece.

with seats and shelters; a show place, accessible by car, with the usual accompaniments of notice boards and baskets for litter.

When Sidmouth Council purchased the greater part of the land on the eastern side of Salcombe Valley to prevent the establishment of a holiday camp there, Cornish gave a good deal of his land on the western side of the valley to the National Trust to ensure that its natural beauty should be preserved in perpetuity. He also created provision for public rights of way.

Cornish (1942) was hugely proud of his family, who had been minor gentry of the Sidmouth district since the fifteenth century. He died in Camberley on 1 May 1948, and was buried alongside his first wife, Ellen, in Salcombe Regis churchyard. In the following year, national parks, for which he had campaigned for so long, were established in England. In 2001, his preserved cliff top became part of the Jurassic Coast World Heritage Site.

John Green

John Frederick Norman Green (1873–1949) was an important geologist who won the Lyell Medal of the Geological Society in 1925, and served as its President from 1934 to 1936. He was also President of the Geologists' Association.

Green was born in Norfolk to the Reverend William Frederick Green and Florence Agnes Green (nee Coles) who came from the former port of East Budleigh, which perhaps explains his work on the plateau surfaces of the East Devon coast. He was educated at Bradfield College in Wiltshire and at Emmanuel College in Cambridge. He died in Bournemouth. He was an amateur, in the sense that most of his working life was spent in the colonial service. He retired from this in 1933 and then devoted himself to geology.

Green was interested in the evolution of the landscape as recorded in the terraces and erosion surfaces of the south west of England, including Dartmoor, East Devon (Figure 7.9), and the coast and raised beaches of southern England, particularly at Bournemouth. He also worked in the Lake District, the Scottish Highlands, Pembrokeshire, Ingleton, and the Dudden Estuary. Green was a truly innovative thinker and tried to relate the terrace sequences to the climate and sea level data established in the Mediterranean. He also noted the Paleogene deposits of the plateau surfaces. He was very much a member of the group of geomorphologists called

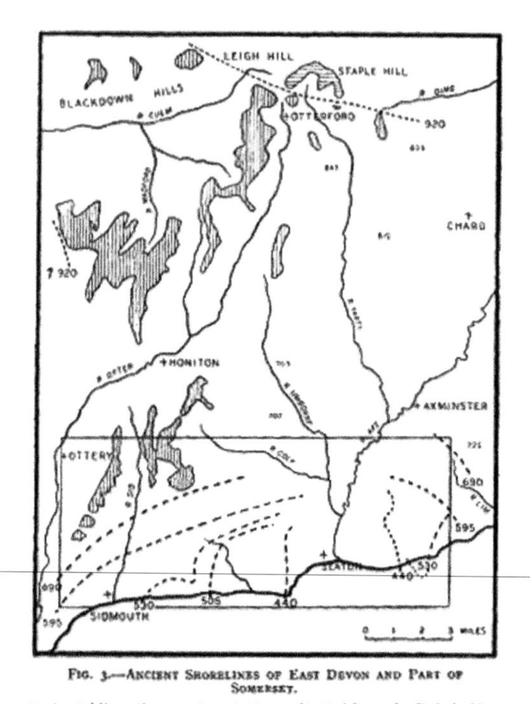

Figure 7.9 Green's ancient shorelines of East Devon, showing the former shorelines inland from Sidmouth and Seaton (Green 1941, figure 3).

With permission of Elsevier.

denudation chronologists, who tried to establish landscape history through the study of erosional and depositional terraces. He was an admirer of Sidney W. Wooldridge. His 1941 paper on the high platforms inland from Sidmouth and the deposits associated with them, including sarsens, is particularly important with respect to trying to establish the long-term evolution of the coastline back in the Cenozoic.

Green's obituary by E. E. S. B. (probably Edmund Ernest Stockwell Brown) for the Geological Society in 1949 hints at Green's character.

> On many subjects Green held decided views, which he would advance strongly, yet persuasively. Sometimes they were unorthodox but often right.

When, occasionally, they were unacceptable to his colleagues, his powers of leadership and advocacy made them the more difficult to combat. If, inadvertently, he thought he had given offence, he took immense pains to make amends. In his geological controversies Green was never dogmatic in the exposition of his views and he maintained the friendliest relations with his opponents. He was always scrupulous in acknowledging his indebtedness to the maps and memoirs of the Geological Survey and to his discussions with, and to the work of, contemporary geologists. It is not easy to summarize the life of an outstanding personality whose interests were so wide and whose work was so diverse. His enthusiasm for each new problem tackled, his physical and mental vigour, the stimulation of the original work he achieved, his warm friendship and kindly help to all who sought it, are a few of the qualities which remain in honoured memory.

Ernest Burton

Ernest St John Burton (1875–1962) was born in Parkstone, Dorset, and was the son of Thomas Arthur Burton, a noted organist and composer. Burton himself became a talented musician and played violin and piano for some years in the Bournemouth Municipal Orchestra. He was also a noted artist. He specialised in painting landscapes, particularly of southern and western England. His works were exhibited in venues that included the Paris Salon, the Royal Institute of Water Colour Painters, and the Royal West of England Academy. However, his hobby was natural history, and he became a Fellow of the Geological Society in 1923, of the Zoological Society in 1924, and of the Linnean Society in 1927. His forte proved to be geology, and this remarkable polymath wrote a series of papers between 1925 and the Second World War. Some of these were on the Eocene deposits at Barton. He lived there from 1934 and died aged 87 years.

Burton's paper (1937) on Lulworth Cove is of particular interest. He wrote (p. 377),

> The well-known Lulworth Cove, situated on the Dorsetshire coast between Mupe Bay, on the east, and Dungy Head, on the west, is generally considered to be a very good example of the result of marine erosion acting upon rocks of unequal resistance, and such indeed it is.

He also reproduced the model of coastal evolution that had been proposed by Aubrey Strahan, which suggested that there was a suite of features that could be employed to show a sequence of coastal evolution (p. 378).

> The first of these examples (Stair Hole), shows the breaching of the Portland Stone in three places, and the consequent washing away of the Purbeck and Wealden Beds. The second (Lulworth Cove) demonstrates a more advanced stage, where the sea has made a breach wide enough to enable a direct attack to be instituted, and a more comprehensive removal of the softer beds within to take place. The third example (Worbarrow Bay), is given to mark a still more advanced condition, resulting in the formation of a wide bay by the almost entire removal of the Portland Stone and Purbeck Beds and all the Wealden Beds, originally extending in unbroken succession across Worbarrow Bay, to the coast opposite at Mupe Rocks, a distance of over one mile and a half.

However, Burton pointed out something that had hitherto been missed and that is the role of spring action and rivers in helping to create these landforms. Lulworth itself, for example, lies at the seaward end of two rather large valley systems, and much of its excavation was due to these streams (Figure 7.10). This interpretation is discussed by Goudie and Brunsden (1997).

In another paper (1932), Burton presented evidence that the Isle of Purbeck, and particularly the upland from Durlston to Worth Matravers, had once been planed off by erosional processes to give what geomorphologists call a *peneplain*.

Marie Stopes

Marie Stopes (1880–1958) (Figure 7.11) is best known as a supporter of womens' rights and as an advocate of birth control. However, she was also a very important palaeobotanist. She was born in Edinburgh, the elder daughter of Henry Stopes (*c.* 1852–1902), an architect from a wealthy brewing family who was interested in archaeology, and his wife, Charlotte (1840–1929), a Shakespearian scholar and promoter of women's education.

Influenced by her father's scientific interests from an early age, when she helped him wash and catalogue hundreds of flints, Stopes chose to take a degree in science at University College in London. She distinguished herself

Figure 7.10 Lulworth Cove, looking west. A large valley system enters the cove from behind the buildings in the rear and may have excavated a large proportion of the total volume of the cove (ASG).

Figure 7.11 Marie Stopes in the laboratory.

Source: https://en.wikipedia.org/wiki/Marie_Stopes#/media/File:Marie_Stopes_in_her_laboratory,_1904.jpg.

during her university career both academically—winning the Gold Medal in botany in her first year—and socially, as President of the Women's Debating Society, where she introduced events in which both sexes took part. In 1902, Stopes achieved her goal of obtaining the BSc in two years with double honours—a first-class degree in botany and a third in geology.

Stopes went on to a remarkably successful scientific career in palaeobotany. She obtained her PhD on Carboniferous flora from Munich University. In 1904, she was appointed assistant lecturer in botany at Manchester University, the first appointee to such a post, and, in 1905, she became the youngest doctor of science in Britain. She was invited to prepare the catalogue of Cretaceous flora for the geological department of the British Museum, participated in international discussions, went down coal mines in pursuit of fossil plants, and wrote a textbook on plant life for young people. In 1911, she even tried to accompany Scott to Antarctica to collect fossils of *Glossopteris*. She worked on a range of palaeobotanical problems, including the origin of flowering plants, the role of plants in the formation of coal, and on the ecology of the coal-forming environments. The great significance of this research has been well described by a leading palaeobotanist of the modern era, Bill Chaloner (2005).

After a series of highly unsatisfactory romantic liaisons, she published *Married Love*, in 1918. This became a best-seller. Later in the year, *Wise Parenthood* appeared, dealing explicitly with contraception. Early in 1921, Stopes and her husband Humphrey Roe inaugurated a birth control clinic in Upper Holloway, North London, which was then a poor working-class area. This was staffed by trained nurses with a female doctor available for cases requiring medical examination. Stopes, a proponent of the study of eugenics, held views on race and the lower social strata that many now find highly unpalatable.

In 1923, she purchased the Old Higher Lighthouse on the Isle of Portland (Figure 7.12), and for many years she and her husband used it as a summer retreat. Stopes's only child, Harry, was, it is hinted, conceived beneath the stars in its lantern tower, and, following his birth in 1924, Stopes largely turned her back on serious geological research. That said, having a base on Portland meant that she could never entirely ignore fossil plants or other geological phenomena. In 1951, in the Bumpers Lane Quarry behind Portland Museum, a large solution-widened fissure became exposed in the quarry face. In it quarrymen found human skulls and animal bones, and these were the subject of one of her last papers (1952).

Figure 7.12 The old Higher Lighthouse, Portland, where Stopes used to live. The lantern tower is in the centre (ASG).

Stopes died from breast cancer, and her remains were deposited in the sea off Portland Bill (Figure 7.13). She is remembered on the Isle for her establishment of Portland Museum (Figure 7.14), which in 2020 installed a plaque to mark both her generosity and her eminence (Figure 7.15). The museum also houses a display devoted to her remarkable life.

Alan Carr

One of the finest scientists to work on the geomorphological processes of the Dorset coast was Alan Paul Carr (1930–2000). His life, together with a full bibliography of his works is provided by Whittaker (2003).

Carr was a geomorphologist who was one of the first, along with the late Clarence Kidson, to become a professional geomorphologist. After graduating in geography at University College London, he spent a short period in teaching. He then worked for the Nature Conservancy, UK, stationed at Furzebrook Research Station, Wareham, Dorset, and later for the Institute of Oceanographic Sciences in Taunton. He worked on beaches and sediment transfers, especially on Orford Ness and Chesil Beach. His seminal papers on Chesil are now the baseline for all our work.

Figure 7.13 Portland Bill at sunset. Marie Stopes's remains were deposited there (ASG).

Figure 7.14 The Portland Museum: Marie Stopes's gift to the Island and Royal Manor of Portland in 1930 (ASG).

Figure 7.15 A plaque on the wall of Portland Museum.
Photo by Professor Heather Viles.

Carr employed pensioners to measure pebbles on Chesil Beach in an attempt to characterise the famous pebble grading patterns, which he found to be more complex than often thought. He also used student workers David Jones and Edward Relph, inspiring both to go on to geographical research and teaching posts at the London School of Economics and the University of York in Canada. Both were students at King's College, London, at the time. He used tracers to measure rates of pebble movement, determined the rates of crest recession and lowering, and employed borehole records, provided mainly by the Central Electricity Generating Board who had plans for a power station in the area, to gain an idea of the beach's evolution and substrates.

Carr was also adept in using old maps and employed them, *inter alia*, in reconstructing both the history of Chesil and the history of the South Haven Peninsula of Studland. His knowledge was in demand by those concerned with coastal defence structures and coastal preservation, and he wrote reports for both Wessex Water and the Dorset County Council. He advised extreme caution in the implementation of any engineering works.

Figure 7.16 Chesil Beach, the Fleet, and Portland Harbour from the Portland Heights (ASG). Chesil was the subject of study by, amongst others, John Coode, Vaughan Cornish, Joseph Prestwich, and Alan Carr.

His painstaking work on Chesil Beach (Figure 7.16) has established a standard for coastal sedimentary environments and a detailed analysis of pebble types and origins, and a good model for the evolution of the beach itself. This is quoted in all recent publications in the scientific and planning, conservation literature.

Carr needs to be remembered, and it is worth noting that the Fleet Study Group, of which he was Chairman from 1997 to 2000 and who regularly carry out surveys of the beach condition and ecosystem, established a strong network of survey location stations on a permanent grid base from which all future surveys will be based. The survey included full global positioning systems (GPS)-controlled baselines and stations on the beach at 500-metre intervals. It is currently used to support beach and lagoon surveys; physical, ecological, and archaeological work; ground control for aerial photography; light detection and ranging (LiDAR); and biological profiles. The wider purposes were to ensure the effective management of the beach and lagoon and to provide an appropriate and lasting memorial to a magnificent scien tist. This is a real legacy—the A. P. Carr Survey network. This is the loose name for the Chesil Beach Survey Stations made and regularly resurveyed by the Fleet Study Group (http://www.chesilbeach.org/FSG/points.pdf). The Fleet's Holocene history has been reconstructed through coring of sediments from its floor by Dr E. D. K. Coombe, an Oxford graduate, a former Portland quarry owner, and brave RAF pilot, who died in 2020 (Coombe 1998, 2000).

When Carr retired, he took up the hobby of bookbinding, at which he became a stylish master; gardening; woodwork; pottery; and learning German. He was a gentle, caring, modest, quiet man who was much loved.

We have as a matter of policy decided not to provide entries on living scientists who have worked on the Jurassic Coast. However, at the suggestion of the Jurassic Coast Trust and some of his friends, Andrew Goudie decided to include a summation of Denys Brunsden, who played such a major and significant role in the designation of the World Heritage Site. Box 7.1 is an appreciation of Denys, Box 7.2 summarises his vision for the World Heritage Site, and Box 7.3 lists his key publications on the Jurassic coast.

Box 7.1 Denys Brunsden: An Appreciation

Professor Denys Brunsden (Figure 7.17) is, it has been said, a man with a huge frame and a huge intellect. He was born in Devon in 1936, played much rugby football at school, and did National Service in the RAF, where he learnt the great skill of air photo interpretation, which was to underpin much of his subsequent career. From the RAF, he went to read geography at King's College London in 1956, and he remained there until he retired four decades later. He undertook doctoral research under Professor Sidney Wooldridge, the great denudation chronologist, on his native Devon, and he became a noted authority on Dartmoor and the River Dart. A formative event occurred, however, when he left the arthritic, sod-covered British landscape for a visit to New Zealand in 1965. Here he was exposed to a very different environment with enormous geomorphological activity. He also gained a great deal from a spell in Louisiana, including his friendship with 'Jess' Walker, a noted coastal geomorphologist.

Denys became a major figure in the study of mass movements, and, with the likes of Professor John Thornes, made significant contributions to the

Figure 7.17 Professor Denys Brunsden with his co-author's daughter, Amy (ASG).

study of fundamental concepts such as landscape sensitivity and change. He played a major role in the creation of applied geomorphology, and he has also been an institution builder. For example, he was involved with the British Geomorphological Research Group (now the British Society for Geomorphology) from its earliest days and became its Chairman. He also helped to establish the International Association of Geomorphologists (IAG) and was its first President (1989–1993). The IAG created an award in honour of Denys's career—the Brunsden Medal—and made him an Honorary Fellow in 1997. He was also President of the Geographical Association in 1986, a post which reflected his great interest in teaching at school level. He remains connected with the World Heritage Site as the first Patron of the Jurassic Coast Trust, which manages the site today. It remains the only natural World Heritage Site in England and Wales. Queen Elizabeth II appointed him OBE for services to geoconservation and geomorphology on 31 December 2003.

Denys has inevitably received various other major awards. He received the Gill Award from the Royal Geographical Society in 1977, was made an Honors Recipient of the Association of American Geographers in 1991, received the David Linton Award of the British Geomorphological Research Group in 1993, and the William Smith Medal from the Geological Society of London (awarded for outstanding research in applied or economic geology) in 2000. In 2001, he presented the Glossop Lecture and, in 2010, the Geological Society awarded him the R. H. Worth Prize. Although a Devonian, he was the recipient of the Society of Dorset Men's first Dorset Award in 2016.

Denys has made major research contributions to the geomorphology of the Dorset coast, and in particular to the study of Black Ven and other mass movement complexes, including those on Portland. He has also co-authored the Geographical Association guides to the classic landforms of Dorset. A selection of some of the papers he has written on the coast are listed at the end of this appreciation.

Box 7.2 A Vision for the World Heritage Site

Denys himself has provided the following background to his vision of the World Heritage Site:

> My journey into the subject of this book is rooted in the training I received from Mr Fred Dutton, a fine geography teacher at Torquay Grammar School; my trade in the RAF as an Aerial Photograph Interpreter, which gave a different perspective on the special places of Earth; and my degree in Geography at King's College London.
>
> This was followed by research and teaching at King's for my whole academic life. Sabbatical visits to New Zealand, Louisiana, Heidelberg; experience setting up the International Association of Geomorphologists; and helping to create a consultancy company called Geomorphological Services Limited, which worked all over the world, gave me huge experience of the beautiful places of Earth and different attitudes to the problems they face as they are subjected to immense development and population pressures.
>
> These experiences invoked deep feelings about the need to look after the precious planet on which we live. Yet I knew that there is only a small

amount one person can do in this vital requirement. My chance to contribute came when I retired early to Dorset. I was asked to give a lecture to a meeting where the development of a Coastal Forum (The Lyme Bay Forum) was being discussed. I described the coast from Portland Bill to Start Point. At the end of the talk I said: 'Anywhere else in the world this level of unique scientific value, wonderful fossils recording the history of life over 190 million years of Earth History, spectacularly important landforms and dynamic processes, plus the fact that it was one of the birth-places of geological science with over 200 years of scientific discovery by dozens of the most brilliant scientists who ever lived, would be declared as a natural World Heritage Site.'

This message was picked up by the chief planning officers of the Counties of Devon and Dorset and eventually we were inscribed on the World Heritage List by the United Nations Educational, Scientific and Cultural Organization (UNESCO) in Helsinki in 2001.

My own research on the landslides and coastal erosion of Dorset over 40 years had given me a deep knowledge and love of the landscape, so that when I became the first Chairman of the Dorset Coast Forum I was able to help draw together many diverse views on the management of the coast. Thus, when the Jurassic Coast project was proposed, there was unanimous agreement and cooperation which, in time, allowed us to develop a comprehensive strategy for its management.

The counties began by developing visions based on key concepts—we wanted to develop personal obligations from each stakeholder group, to reconcile views, integrate actions, and to determine statutory obligations for administrative bodies. Central to the consensus view was the idea that planning regulation could often be replaced by voluntary initiative. Understanding the interest and knowledge of each Community Group was seen to be the way forward if we could be encouraged to work sympathetically together.

This was founded on the belief that the coast was the heart and lungs of the nation, to be looked after and made accessible for all. Geodiversity was regarded as a community asset. Risk and pollution could be assessed, monitored, and managed. These decisions could enhance the sustainable quality of life of inhabitants and visitors.

Underlying this was the idea that knowledge and education were the key to all our actions. Decisions were to be based on the best social, scientific, and economic information. I have written many times that 'We developed a belief that education became powerful because, as people came to understand the coast, they came to respect it; respect meant they valued it and that led to a sense of personal ownership. Because they felt they had a personal stake they wanted to care for it. Legislation was not needed; public opinion and attitude became very strong. The role of the Forum and, later, the Dorset and Devon World Heritage Site management team thus became coordination, stimulation, facilitation.'

The wider vision that emerged was that our Earth Science assets should be recorded and for decisions to be based on the best information. We wanted this information to be made accessible to a wide range of users and future generations through education, art and scientific programmes, museums and publications; to inform, stimulate, and enhance people's lives and be a catalyst for the regeneration of places and communities. We wanted to create a sense of pride in the community and develop a ground-breaking education programme for both the public and science.

It quickly emerged that, central to this, was the belief that the Jurassic Coast was one of the most significant Earth Science sites in the world, displaying a remarkable set of geological, palaeontological, and geomorphological features and possessing a unique historical importance in the founding of geology, palaeontology, and geomorphology. This was my motive in contributing to this volume because I feel that all of us can be enriched by a strong sense of time and place.

Box 7.3 Research Work on the Jurassic Coast World Heritage Site

The following publications give a flavour of the research work that Denys has undertaken on the Jurassic Coast World Heritage Site over a period of almost half a century:

Brunsden D., and Jones, D. K. C. 1972. The morphology of degraded landslide slopes in South West Dorset. *Quarterly Journal of Engineering Geology and Hydrogeology* 5, 205–222.

Brunsden D., and Jones, D. K. C. 1976. The evolution of landslide slopes in Dorset. *Philosophical Transactions of the Royal Society of London* A (283), 605–631.

Pitts, J., and Brunsden, D. 1987. A reconsideration of the Bindon landslide of 1839. *Proceedings of the Geologists' Association* 98, 1–18.

Chandler, J. H., and Brunsden, D. 1995. Steady state behaviour of the Black Ven mudslide: The application of archival analytical photogrammetry to studies of landform change. *Earth Surface Processes and Landforms* 20, 255–275.

Brunsden, D. 1996. Landslides of the Dorset coast: Some unresolved questions. *Proceedings of the Ussher Society* 9, 1–11.

Brunsden, D., and Chandler, J. H. 1996. Development of an episodic landform change model based upon the Black Ven mudslide, 1946–1995. In M. G. Anderson and S. M. Brooks (eds.), *Advances in Hillslope Processes* 2, 869–896. Chichester: Wiley–Blackwell.

Brunsden, D., Coombe, K., Goudie, A. S., and Parker, A. G. 1996. The structural geomorphology of the Isle of Portland, southern England. *Proceedings of the Geologists' Association* 107, 209–230.

Brunsden, D., and Moore, R. 1999. Engineering geomorphology on the coast: Lessons from West Dorset. *Geomorphology* 31, 391–409.

Lee, E. M., and Brunsden, D. 2001. Sediment budget analysis for coastal management, West Dorset. *Geological Society London, Engineering Geology Special Publications* 18, 181–187.

Brunsden, D., and Badman, T. (eds.). 2003. *The Official Guide to the Jurassic Coast, Dorset and East Devon's World Heritage Coast: A Walk Through Time*. Wareham: Coastal Publications.

Conclusion

In this book we have concentrated on discussing the past. We have described the lives and achievements of some remarkable people who are now deceased. However, we are very conscious that much excellent work continues today, involving a large number of geologists, palaeontologists and geomorphologists. We are grateful to many of them for their advice, knowledge and stimulation. It is encouraging that so much work is still being done on the rocks and landscapes of this outstanding World Heritage Site, and that re-evaluations of the pioneers are being undertaken. It is clear that for over three hundred years this stretch of coastline has been absolutely seminal in the development of many of the greatest ideas that underlie the evolution of the Earth sciences, and has thus had an impact on scientific thought all over the world. It has moulded our views on the extinction and evolution of species, the nature of the stratigraphic record, the origin of sediments, and the evolution of landforms. The role of the Site in the "Development of the History of Science" was the third criterion used in the designation of the Dorset and East Devon Coast as a World Heritage site.

Geological Pioneers of the Jurassic Coast. Andrew S. Goudie and Denys Brunsden, Oxford University Press.
 DOI: 10.1093/oso/9780197638088.003.0008

Glossary

ammonite Extinct cephalopods especially abundant in the Mesozoic age that had flat spiral shells with the interior divided by septa into chambers.

belemnite An extinct cephalopod mollusc allied to the cuttlefish with a bullet-shaped internal shell that is typically found as a fossil in marine deposits of the Jurassic and Cretaceous periods.

biostratigraphy The branch of stratigraphy concerned with fossils and their use in dating rock formations.

blockstream A linear body of boulders or blocks of rock formed by movement down a slope or valley.

catastrophism The belief that major changes in the Earth's history result from sudden events (e.g., Noah's Flood) rather than through gentle change.

Cenozoic Third of the major eras of Earth's history, beginning about 66 million years ago and extending to the present. It is generally divided into three periods: the Paleogene (66 million to 23 million years ago), the Neogene (23 million years ago), and the Quaternary (2.6 million years ago).

cephalopod A class of molluscs with distinct tentacle heads, which includes both fossil ammonites and the modern squids, octopuses, and nautilus.

chert A type of sedimentary rock composed of microcrystalline or cryptocrystalline quartz, the mineral form of silicon dioxide (SiO_2). It occurs as nodules, concretionary masses, and as layered deposits. It breaks with a conchoidal fracture, often producing exceptionally sharp edges.

coprolites Fossilised faeces of animals.

Coral Rag A rubbly, shelly limestone (some of which is composed of corals) from the top of the Upper Jurassic Corallian Group.

Cornbrash A Middle Jurassic geological formation. It ranges in age from Bathonian to Callovian, the uppermost part of the Middle Jurassic. It is often composed of fossiliferous limestone.

Cretaceous Relating to or denoting the last period of the Mesozoic era, between the Jurassic and Tertiary periods, *c.* 145 to 66 million years ago.

denudation chronology The branch of geomorphology that deals with the historical development of landscapes by denudation. Evidence for stages of landform development is provided by studies of erosion surfaces (such as peneplains) and the deposits that mantle them, stream long-profiles, and drainage patterns.

diluvial Relating to, or brought about by a flood (e.g., Noah's Flood).

dinosaur Any of a group (Dinosauria) of extinct, often very large, chiefly terrestrial carnivorous or herbivorous reptiles of the Mesozoic era.

doline A closed depression formed in limestone areas as a result of the solution of the rock.

dry valley A valley formerly occupied by a permanent or seasonal stream but which now seldom or never experiences flow.

Eocene An epoch of the Tertiary period, occurring from 55 to 40 million years ago and characterised by the advent of the modern mammalian orders.

facies The characteristics of a rock or sediment unit that reflect its environment of deposition and allow it to be distinguished from rock or sediment deposited in an adjacent environment.

flint A hard, sedimentary cryptocrystalline form of the mineral quartz categorised as the variety of chert that occurs in chalk or marly limestone.

Forest Marble A mix of mudstones and limestones that dates back to the late Bathonian stage of the Middle Jurassic.

geomorphology The study of landforms and the processes that form them.

Glacial theory A theory, developed in the 1830s and 1840s that there had been an Ice Age and that most of North America, northern Europe, and the north of Asia had been covered in ice sheets during a period called the Pleistocene. It was an hypothesis that was used to explain erosion and the subsequent deposition of boulder clay (till). It helped to explain the development of landforms in areas that are not now glaciated.

ichthyosaur An extinct marine reptile of the Mesozoic era resembling a dolphin, with a long pointed head, four flippers, and a vertical tail. It became extinct in the Middle Cretaceous.

Jurassic Relating to or denoting the second period of the Mesozoic era, c. 200 to 145 million years ago, between the Triassic and Cretaceous periods.

Liassic The lowest series of rocks of the Jurassic system, dating from c. 200 to 180 million years ago.

lithology A description of the major macroscopic features of a type of rock, such as texture, composition, and colour.

mass movement The movement of material down a slope under the influence of gravity. It includes such phenomena as mudflows and landslides.

Mesozoic Relating to or denoting the era between the Palaeozoic and Cenozoic eras, comprising the Triassic, Jurassic, and Cretaceous periods.

midden A refuse heap.

mudflow A flowing mass of soft, wet, unconsolidated earth and fine-grained debris made fluid by rain or melted snow and often building up great speed.

palaeobotany The branch of palaeontology or palaeobiology dealing with the recovery and identification of plant remains from geological contexts and their use for the biological reconstruction of past environments.

palaeoecology The study of fossil animals and plants in order to deduce their ecology and the environmental conditions in which they lived.

palaeontology The branch of science concerned with fossil animals and plants.

palaeosol A fossil soil.

Paleogene The Palaeogene spans 43 million years from the end of the Cretaceous period (66 million years ago) to the beginning of the Neogene period (23 million years ago). It is the beginning of the Cenozoic era.

peneplain Gently undulating, almost featureless plain that, in principle, would be produced by fluvial erosion that would, in the course of geologic time, reduce the land almost to baselevel (sea level), leaving so little gradient that essentially no more erosion could take place.

Pleistocene The geological epoch that lasted from *c.* 2.6 million years ago until c. 11,700 years ago. It saw Earth's most recent period of repeated glaciations.

plesiosaur A large extinct marine reptile of the Mesozoic era, with a broad flat body, large paddle-like limbs, and typically a long flexible neck and small head. Buckland likened it to 'a snake strung through a turtle.'

pterodactyl Any of a number of genera of flying reptiles of the extinct order Pterosauria, from the Jurassic and Cretaceous periods, having a highly reduced tail, teeth, and a bird-like beak.

pterosaur A fossil flying reptile of the Jurassic and Cretaceous periods, with membranous wings supported by a greatly lengthened fourth finger.

Quaternary The current and most recent of the three periods of the Cenozoic era in the geologic time scale. The Quaternary period is divided into two epochs: the Pleistocene (*c.* 2.6 million years ago to 11.7 thousand years ago) and the Holocene (11.7 thousand years ago to today).

rotational slip A variety of landslide characterised by movement along a concave-up failure surface. The upper unit of the slump is typically tilted back, and surface water may be retained in the depressed zone.

sabkha A hyper-saline coastal marsh of the type that occurs on the desert coast of the Arabian Gulf.

sarsen A silicified sandstone boulder of a kind which occurs on the chalk downs of southern England. Probably remnants of eroded Tertiary beds.

saurian Any of a suborder (Sauria) of reptiles, including the lizards and, in older classifications, the crocodiles and various extinct forms (such as the dinosaurs and ichthyosaurs) that resemble lizards.

stratigraphy The branch of geology concerned with the order and relative position of strata (layers of rock distinguishable from those above and below) and their relationship to the geological time scale.

stratum (plural strata) A defined layer, or layers, of sedimentary rock, usually separated from other beds both above and below by bedding planes.

Tertiary A largely obsolete term for one of the great and more recent of the subdivisions of the geological time scale, following the Cretaceous and including the Eocene, Oligocene, Miocene, and Pliocene.

thrombolites Clotted accretionary structures formed in shallow water by the trapping, binding, and cementation of sedimentary grains by biofilms produced by micro-organisms, including cyanobacteria.

translational slide A down-slope movement of material that occurs along a distinctive surface of weakness, such as a fault, joint, or bedding plane. If this slip surface is approximately straight, then it is termed translational.

Triassic Relating to or denoting the earliest period of the Mesozoic era, between the Permian and Jurassic periods, occurring from *c.* 250 to 200 million years ago and characterised by the advent of dinosaurs, pterosaurs, plesiosaurs, ichthyosaurs, etc.

tufa A porous rock composed of calcium carbonate and formed by precipitation from water (e.g., around mineral springs, in stream beds, and at waterfalls).

unconformity The surface of contact between two series of rocks that is sufficiently unconformable that it implies the passage of time and often the occurrence of Earth movement and erosion between two periods of deposition.

valley bulge An upward arching of the bedrock along the axis of a valley. It may not be visible at the ground surface due to subsequent erosion, but it is revealed by the distortion of the geologic structure. It may be due to frost heave or to the compressive forces set up by the weight of overlying strata causing extrusion of clays into an incised valley floor.

Bibliography and References

This section contains both material referred to in the text and a bibliography of publications on the Jurassic Coast's geology and geomorphology. It starts with a list of general publications relating to Chapter 1 and to the introductions to the following chapters. This is followed by publications related to the individual scientists treated in this book. These are arranged alphabetically.

General

Ager, D. V., and Smith, W. E. 1965. *The Coast of South Devon and Dorset Between Branscombe and Burton Bradstock*. Geologists' Association Geological Guides No. 23. Geologists' Association, London.

Allison, R. J. 2020. The landslides of East Devon and West Dorset. In A. S. Goudie and P. Migoń (eds.), *Landscapes and Landforms of England and Wales*. Springer, Cham, 201–213.

Bailey, E. 1962. *Charles Lyell*. Thomas Nelson and Sons, London.

Barton, C. M., and 13 others. 2011. *Geology of South Dorset and South-East Devon and Its World Heritage Coast*. Special Memoir of the British Geological Survey. Keyworth, Nottingham.

Benton, M. J., and Spencer, P. S. 1995. *Fossil Reptiles of Great Britain*. Joint Nature Conservation Committee, Geological Review Series, 10. London, Chapman & Hall.

Berger, J. F. 1811. A sketch of the geology of some parts of Hampshire and Dorsetshire. *Transactions of the Geological Society of London*, 1, 249–268.

Bettany, G. T. 2004. Egerton, Sir Philip de Malpas Grey, tenth baronet (1806–1881). Oxford Dictionary of National Biography (https://www.oxforddnb.com/view/10.1093/ref:odnb/9780198614128.001.0001/odnb-9780198614128-e-8591).

Bosence, D., and Gallois, A. 2021. How do thrombolites form? Multiphase construction of lacustrine microbialites, Purbeck Limestone Group (Jurassic), Dorset, UK. *Sedimentology*. doi:10.1111/sed.12933.

Bray, M. J. 1996. *Beach Budget Analysis and Shingle Transport Dynamics in West Dorset*. Unpublished doctoral dissertation, London School of Economics and Political Science.

Brunsden, D. (ed.). 2003. *The Official Guide to the Jurassic Coast: Dorset and East Devon's World Heritage Coast: A Walk Through Time*. Coastal Publishing, Wareham.

Brunsden, D. 2019. The Dorset and East Devon Coast Heritage Site: A vision. *Proceedings of the Geologists' Association*, 130, 263–264.

Brunsden, D., and Edmonds, R. 2009. The Dorset and East Devon Coast: England's Geomorphological World Heritage Site. In P. Migoń (ed.), *Geomorphological Landscapes of the World*. Springer, Dordrecht, 211–221.

Brunsden, D., and Goudie, A. 1997. *Classic Landforms of the West Dorset Coast*. Geographical Association, in conjunction with the British Geomorphological Research Group.

Burek, C. V., and Higgs, B. 2007. The role of women in the history and development of geology: An introduction. *Geological Society, London, Special Publications*, 281(1), 1–8.

Butler-Warke, A., and Warke, M. 2021. Foundation stone of empire: The role of Portland Stone in 'heritage', commemoration, and identity. *Transactions of the Institute of British Geographers*. doi:10.1111/tran.12462.

Cadbury, D, 2000. *The Dinosaur Hunters*. Fourth Estate, London.

Campbell, D. 2006. *Exploring the Undercliffs*. Coastal Publishing, Wareham.

Chorley, R. J., Dunn, A. J., and Beckinsale, R. P. 1964. *The History of the Study of Landforms or the Development of Geomorphology. Volume 1. Geomorphology Before Davis*. London, Methuen.

Cope, J. C. W. 2016. *Geology of the Dorset Coast*. Geologists' Association, London.

Coram, R. A., and Radley, J. D. 2021. Revisiting climate change and palaeoenvironments in the Purbeck Limestone Group (Tithonian–Berriasian) of Durlston Bay, southern UK. *Proceedings of the Geologists' Association*, 132, 392–404.

Coram, R. A., Radley, J. D., and Benton, M. J. 2019. The Middle Triassic (Anisian) Otter Sandstone biota (Devon, UK): Review, recent discoveries and ways ahead. *Proceedings of the Geologists' Association*, 130, 294–306.

Darwin, C. 1859. *On the Origin of Species*. John Murray, London.

Davies, G. M. 1956. *The Dorset Coast: A Geological Guide*. A. and C. Black, London.

Dineley, D. L., and Metcalf, S. J. 1999. *Fossil Fishes of Great Britain*. Joint Nature Conservation Committee, Peterborough.

Dorset County Council. 2000. *Nomination of the Dorset and East Devon Coast for inclusion in the World Heritage* List. Dorset County Council, Dorchester.

Duffin, C. J. 2010. Brief history of palaeontological work. In A. R. Lord and P. G. Davis (eds.), *Fossils from the Lower Lias of the Dorset Coast*. Palaeontological Association, London, 8–25.

Dykes, A. P., and Bromhead, E. N. 2021. The Southwell Topple: Reassessment of a very large coastal toppling failure on the Isle of Portland, UK. *Quarterly Journal of Engineering Geology and Hydrogeology*, doi:10.1144/qjegh2020-146.

Edmonds, R. 2019. High resolution mapping, description and interpretation of the foreshore in front of the great undercliff landslides; the Plateau, Dowlands and Bindon within the Axmouth to Lyme Regis Undercliffs National Nature Reserve, south Devon coast, England, and how that might inform understanding of the mechanisms of failure. *Proceedings of the Geologists' Association*, 130(3–4), 473–482.

Edwards, R. A. 2008. *The Red Coast Revealed: Exmouth to Lyme Regis*. Coastal Publishing, Wareham.

Egerton, P. 1876. Notice of Harpactes velox, a predaceous ganoid fish of a new genus, from the Lias of Lyme Regis. *Geological Magazine*, 3(10), 441–442.

Egerton, P. D. M. G. 1858. XXXV. On chondrosteus, an extinct genus of the sturionidæ, found in the lias formation at Lyme Regis. *Philosophical Transactions of the Royal Society of London*, 148, 871–885.

Egerton, P. D. M. G. 1871. On a new Chimæroid fish from the Lias of Lyme Regis (Ischyodus orthorhinus). *Quarterly Journal of the Geological Society*, 27(1–2), 275–279.

Egerton, P. D. M. G. 1872a. On Prognathodus Güntheri (Egerton), a new genus of fossil fish from the Lias of Lyme Regis. *Annals and Magazine of Natural History*, 9(52), 325–325.

Egerton, P. D. M. G. 1872b. On two specimens of Ischyodus from the Lias of Lyme Regis. *Quarterly Journal of the Geological Society*, 28(1–2), 237–237.

Egerton, P. G. 1868. On the characters of some new fossil fish from the Lias of Lyme Regis. *Quarterly Journal of the Geological Society*, 24(1–2), 499–505.

Elder, E. S. 1982. Women in early geology. *Journal of Geological Education*, 30(5), 287–293.

Ensom, P. C. 2007. The Purbeck Limestone Group of Dorset, southern England. *Geology Today*, 23(5), 178–185.

Francis, J. E. 1984. The seasonal environment of the Purbeck (Upper Jurassic) fossil forests. *Palaeogeography, Palaeoclimatology, Palaeoecology*, 48(2–4), 285–307.

Gallois, R. W. 2005. The development and origin of karst in the Upper Greensand Formation (Cretaceous) of south-west England. *Proceedings of the Ussher Society*, 11(1), 30–36.

Gallois, R. W. 2007. A recent landslide on the east Devon coast, UK. *Quarterly Journal of Engineering Geology and Hydrogeology*, 40(1), 29–34.

Gallois, R. W. 2008. Geological controls on the failure mechanisms within the Black Ven-Spittles landslip complex, Lyme Regis, Dorset. *Proceedings of the Ussher Society*, 12(1), 9–14.

Gallois, R. W. 2011. Natural and artificial influences on coastal erosion at Sidmouth, Devon, UK. *Geoscience in South-West England: Proceedings of the Ussher Society*, 12(3), 304–312.

Gallois, R. 2014. Landslide mechanisms in the Axmouth to Lyme Regis Undercliffs National Nature Reserve, Devon, UK. *Geoscience in South-West England*, 13, 345–359.

Gallois, R. 2019. The stratigraphy of the Permo-Triassic rocks of the Dorset and East Devon Coast World Heritage Site, UK. *Proceedings of the Geologists' Association*, 130(3–4), 274–293.

Gallois, A., Bosence, D., and Burgess, P. M. 2018. Brackish to hypersaline facies in lacustrine carbonates: Purbeck Limestone Group, Upper Jurassic–Lower Cretaceous, Wessex Basin, Dorset, UK. *Facies*, 64(2), 1–39.

Gallois, R. W., and Owen, H. G. 2019. The stratigraphy of the Gault and Upper Greensand Formations (Albian stage, Cretaceous) of the Dorset and East Devon Coast World Heritage Site, UK. *Proceedings of the Geologists' Association*, 130(3–4), 390–405.

Godden, M. 2016. Digging deep for Portland stone. *Geology Today*, 32(4), 143–147.

Goudie, A. S. 2020. The geomorphology of the Isle of Portland. In A. S. Goudie and P. Migoń (eds.), *Landscapes and Landforms of England and Wales*. Springer, Cham, 183–199.

Goudie, A. S., and Brunsden, D. 1997. *Classic Landforms of the East Dorset Coast*. Geographical Association in conjunction with the British Geomorphological Research Group, Sheffield.

Goudie, A. S., and Migoń, P. 2020. The eastern coast of Dorset. In A. S. Goudie and P. Migoń (eds.), *Landscapes and Landforms of England and Wales*. Springer, Cham, 169–182.

Hackman, G. 2014. *Stone to Build London*. Folly Books, Monkton Farleigh.

Hart, M. B. 2009. *Dorset and East Devon. Landscape and Geology*. Crowood Press, Marlborough.

Hart, M. B. 2014. The 'Otter Sandstone River 'of the mid-Triassic and its vertebrate fauna. *Proceedings of the Geologists' Association*, 125(5–6), 560–566.

Jones, T. R. 1881. Address at the opening of the session, 1880–81: November 5, 1880. *Proceedings of the Geologists' Association*, 7(1), 1–57.

Kinns, S. 1895. *Moses and Geology: Or the Harmony of the Bible with Science: Thoroughly Revised, and the Astronomical and Other Facts Brought Up to Date*. Cassell, London.

Kong, D. Y., Lim, J. D., and Huh, M. 2009. Characteristics of the Dorset and East Devon Coast, a UNESCO World Natural Heritage Site, United Kingdom. *Journal of Korean Nature*, 2(2), 99–102.

Larwood, J. 2019. The Jurassic Coast: Geoscience and education: An overview. *Proceedings of the Geologists' Association*, 130, 265–273.

Lord, A. R., and Davis, P. G. (eds.). 2010. *Fossils from the Lower Lias of the Dorset Coast*. The Palaeontological Association, London.

MacCulloch, J. A. 1836. *Geological Map of Scotland*. Arrowsmith, London.

Maddox, B. 2017. *Reading the Rocks: How Victorian Geologists Discovered the Secret of Life*. Bloomsbury, London.

May, V. J. 2003a. Lyme Regis to Golden Cap, Dorset. In V. J. May and J. D. Hansom (eds.), *Coastal Geomorphology of Great Britain*. Geological Conservation Review Series. Joint Nature Conservation Committee, 28, 151–158. Joint Nature Conservation Committee, London.

May, V. J. 2003b. Chesil Beach, Dorset. In V. J. May and J. D. Hansom(eds.), *Coastal Geomorphology of Great Britain*. Geological Conservation Review Series. Joint Nature Conservation Committee, 28, 254–265. Joint Nature Conservation Committee, London.

May, V. J. 2019. The submarine landscape of the 'Jurassic Coast' World Heritage Site, Dorset, UK and its setting. *Proceedings of the Geologists' Association*, 130(3–4), 463–472.

McGowan, C. 2001. *The Dragon Seekers: The Discovery of Dinosaurs During the Prelude to Darwin*. Little Brown, London.

Mortimore, R. N. 2019. Late Cretaceous stratigraphy, sediments and structure: Gems of the Dorset and East Devon Coast World Heritage Site (Jurassic Coast), England. *Proceedings of the Geologists' Association*, 130(3–4), 406–450.

Norman, D. B. 2021. Scelidosaurus harrisonii (Dinosauria: Ornithischia) from the Early Jurassic of Dorset, England: Biology and phylogenetic relationships. *Zoological Journal of the Linnean Society*, 191(1), 1–86.

Nowell, D. A. G. 1998. The geology of Lulworth Cove, Dorset. *Geology Today*, 14, 71–74.

Nowell, D. A. G. 2000. The geology of Worbarrow, Dorset. *Geology Today*, 16, 71–76.

Penn, S. J., Sweetman, S. C., Martill, D. M., and Coram, R. A. 2020. The Wessex Formation (Wealden Group, Lower Cretaceous) of Swanage Bay, southern England. *Proceedings of the Geologists' Association*, 131(6), 679–698.

Pitts, J. 1979. Morphological mapping in the Axmouth-Lyme Regis Undercliffs, Devon. *Quarterly Journal of Engineering Geology and Hydrogeology*, 12(3), 205–217.

Pitts, J. 1983. The temporal and spatial development of landslides in the Axmouth-Lyme Regis Undercliffs National Nature Reserve, Devon. *Earth Surface Processes and Landforms*, 8(6), 589–603.

Pitts, J., and Brunsden, D. 1987. A reconsideration of the Bindon landslide of 1839. *Proceedings of the Geologists' Association*, 98(1), 1–18.

Ray, J. 1673. *Observations Topographical, Moral, & Physiological Made in a Journey through Part of the Low-countries, Germany, Italy, and France with a Catalogue of Plants Not Native of England, Found Spontaneously Growing in Those Parts, and Their Virtues*. John Martyn, London.

Rousseau, J.-J. 1762. *Du Contrat Social ou Principes du Droit Politique*. Rey, Amsterdam.

Rudwick, M. J. S. 2008. *Worlds before Adam. The Reconstruction of Geohistory in the Age of Reform*. University of Chicago Press, Chicago and London.
Ruffell, A. H., and Batten, D. J. 1994. Uppermost Wealden facies and Lower Greensand Group (Lower Cretaceous) in Dorset, southern England: Correlation and palaeoenvironment. *Proceedings of the Geologists' Association* 105(1), 53–69.
Simmons, M. S. 2019. *Great Geologists*. https://view.joomag.com/exploration-insights-great-geos-ebook/0172709001539012700?short.
Smith, W. 1815. *A Delineation of the Strata of England and Wales: With Part of Scotland; Exhibiting the Collieries and Mines, the Marshes and Fen Lands . . ., and the Varieties of Soil*. J. Carey, London.
Smith, W. 2021. *STRATA. William Smith's Geological Maps*. Museum of Natural History, Oxford and Thames & Hudson, London.
Torrens, H. 2004. How the long history of geological studies in Dorset confirms its World Heritage Coast status. *Open University Geological Society Journal*, 25(2), 1–16.
von Zittel, K. A. von, 1901. *History of Geology and Palaeontology to the End of the Nineteenth Century; Translated by Maria M. Ogilvie-Gordon*. Walter Scott and Charles Scribner's Sons, London and New York.
Wilson, V., Welch. F. B. A., Robbie, J. A., and Green, G. W. 1958. *Geology of the country around Bridport and Yeovil*. Geological Survey, London.
Winchester, S. 2001. *The Map that Changed the World: The Tale of William Smith and the Birth of a Science*. Viking, London.
Woodward, H. B. 1907. *The History of the Geological Society of London*. Geological Society, London.
Woodward, H. B. 1911. *History of Geology*. Watts and Co., London.

Individuals

Anning

Curwen, E. C. (ed.). 1940. *The Journal of Gideon Mantell*. Oxford University Press, Oxford.
Davis, L. E. 2012. Mary Anning: Princess of palaeontology and geological lioness. *The Compass: Earth Science Journal of Sigma Gamma Epsilon*, 84(1), 57–88.
Lang, W. D. 1939. Mary Anning (1799–1847), and the pioneer geologists of Lyme. *Proceedings of the Dorset Natural History and Archaeological Society*, 60, 142–164.
Owen, R. 1841. *Report on British Fossil Reptiles*. R., and J. E. Taylor, London.
Sharpe, T. 2020. *The Fossil Woman: A Life of Mary Anning*. Dovecote Press, Wimborne Minster.
Sharpe, T. 2021. A case of mistaken identity: Is Mary Anning (1799–1847) actually William Buckland (1784–1856)? *Earth Sciences History*, 40(1), 68–83.
Tickell, C. 1996. *Mary Anning of Lyme Regis*. Philpot Museum, Lyme Regis.
Torrens, H. S. 1995. Mary Anning (1799–1847) of Lyme; 'The greatest fossilist the world ever knew'. *British Journal for the History of Science*, 28(3), 257–284.
Torrens, H. S. 2004. Anning, Mary (1799–1847). *Oxford Dictionary of National Biography*, https://doi.org/10.1093/ref:odnb/568.
Vincent, P., Taquet, P., Fischer, V., Bardet, N., Falconnet, J., and Godefroit, P. 2014. Mary Anning's legacy to French vertebrate palaeontology. *Geological Magazine*, 151(1), 7–20.

Arber

Arber, E. A. N. 1911. *The Coast Scenery of North Devon: Being an Account of the Geological Features of the Coast-line Extending from Porlock in Somerset to Boscastle in North Cornwall.* JM Dent & Sons, Ltd., London.

Arber, M. A. 1941. The coastal landslips of west Dorset. *Proceedings of the Geologists' Association*, 52(3), 273–283.

Arber, M. A. 1946. The valley system of Lyme Regis. *Proceedings of the Geologists' Association*, 57(1), 8–15.

Arber, M. A. 1973. Landslips near Lyme Regis. *Proceedings of the Geologists' Association*, 84(2), 121–133.

Arber, M. A. 1976. The evolution of landslide slopes in Dorset: Discussion. *Philosophical Transactions of the Royal Society of London Series A*, 283, 631.

Robinson, E. 2005. Muriel Agnes Arber, 1913–2004. *Proceedings of the Geologists' Association*, 116, 61–63.

Robinson, E. 2007. The influential Muriel Arber: A personal reflection. *Geological Society, London, Special Publications*, 281(1), 287–294.

Arkell

Arkell, W. J. 1933. *The Jurassic System in Great Britain.* Clarendon Press, Oxford.

Arkell, W. J. 1938. Three tectonic problems of the Lulworth district: Studies on the middle limb of the Purbeck fold. *Quarterly Journal of the Geological Society*, 94, 2–54.

Arkell, W. J. 1947. *The Geology of the Country around Weymouth, Swanage, Corfe and Lulworth.* Geological Survey, London.

Arkell, W. J. 1951. The structure of Spring Bottom Ridge, and the origin of the mud-slides, Osmington, Dorset. *Proceedings of the Geologists' Association*, 62(1), 21–30.

Arkell, W. J. 1956. *Jurassic Geology of the World.* Oliver and Boyd, Edinburgh and London.

Arkell, W. J. 1958. *Seven Poems.* Cambridge University Press, Cambridge.

Brighton, A. G. 1958. Dr. W. J. Arkell, F. R. S. *Nature*, 181, 1373.

Cox, L. R. 1958. William Joscelyn Arkell 1904–1958. *Biographical Memoirs of the Royal Society*, 4, 1–14.

Cox, L. R. (revised by J. H. Callomon). 2004. Arkell, William Joscelyn (1904–1958). *Oxford Dictionary of National Biography*, https://doi.org/10.1093/ref:odnb/30441.

Torrens, H. 2004. How the long history of geological studies in Dorset confirms its World Heritage Coast status. *Open University Geological Society Journal*, 25(2), 1–16.

Beckles

Austen, J. H. 1852. *A Guide to the Geology of the Isle of Purbeck, and the South-west Coast of Hampshire.* W. Shipp, Blandford.

Beckles, S. H. 1851. On supposed casts of footprints in the Wealden. *Quarterly Journal of the Geological Society of London*, 7, 117.

Beckles, S. H. 1852. On the Ornithoidichnites of the Wealden. *Quarterly Journal of the Geological Society*, 8, 396–397.

Beckles, S. H. 1854. On the Ornithoidichnites of the Wealden. *Quarterly Journal of the Geological Society*, 10, 456–464.

Beckles, S. H. 1862. On some natural casts of reptilian footprints in the Wealden Beds of the Isle of Wight and of Swanage. *Quarterly Journal of the Geological Society*, 18, 443–447.

Coram, R. A., and Radley, J. D. 2021. Revisiting climate change and palaeoenvironments in the Purbeck Limestone Group (Tithonian–Berriasian) of Durlston Bay, southern UK. *Proceedings of the Geologists' Association*, 132, 392–404.

Duffin, C. J. 2012. Coprolites and characters in Victorian Britain. *Vertebrate Coprolites. New Mexico Museum of Natural History and Science, Bulletin*, 57, 45–60.

Falconer, H. 1857. Description of two species of the fossil mammalian genus *Plagiaulax* from Purbeck. *Quarterly Journal of the Geological Society*, 13, 261–282.

Kingsley, C. 1857. Geological discoveries at Swanage. *Illustrated London News*, 26 December, 637–638.

Sarjeant, W. A., Delair, J. B., and Lockley, M. G. 1998. The footprints of Iguanodon: A history and taxonomic study. *Ichnos*, 6, 183–202.

Sweetman, S. C., Martill, D. M., and Smith, G. 2018. Notes on the discovery of two Eutherian mammals in the 'Mammal Bed' of the Purbeck Group (Early Cretaceous, Berriasian) exposed in Durlston Bay, Dorset, UK. *Proceedings of the Dorset Natural History and Archaeological Society*, 139, 105–124.

Westwood, J. O. 1854. Contributions to fossil entomology. *Quarterly Journal of the Geological Society*, 10, 378–396.

Bristow

Anon. 1889. Henry William Bristow, F. R. S. *Nature*, 40, 206–207.

Bristow, H. W. 1856. *Comparative Vertical Sections of the Purbeck Strata*. Geological Survey of Great Britain, Nos. 1, 2, and 3 by H. W. Bristow; No. 4 by O. Fisher and H. W. Bristow. Geological Survey of Great Britain, London.

Bristow, H. W., and Whitaker, W. 1869. On the formation of the Chesil Bank, Dorset. *Geological Magazine*, 6(64), 433–438.

Codrington, T. 1870. Some remarks on the formation of the Chesil Bank. *Geological Magazine*, 7(67), 23–25.

Thackray, J. C. 2004. Bristow, Henry William (1817–1889). *Oxford Dictionary of National Biography*, https://doi.org/10.1093/ref:odnb/3451.

Woodward, H. B. 1889. Henry William Bristow, F. R. S., F. G. S., *Geological Magazine*, 6, 381–385.

Brodie

Brodie, P. B. 1845. *A History of the Fossil Insects in the Secondary Rocks of England: Accompanied by a Particular Account of the Strata in which they Occur, and of the Circumstances Connected with their Preservation*. John Van Voorst, London.

Brodie, P. B. 1853a. Notice of the occurrence of an elytron of a coleopterous insect in the Kimmeridge Clay at Ringstead Bay, Dorsetshire. *Quarterly Journal of the Geological Society*, 9(1–2), 51–52.

Brodie, P. B. 1853b. On the occurrence of the remains of insects in the Tertiary clays of Dorsetshire. *Quarterly Journal of the Geological Society*, 9(1–2), 53–54.

Brodie, P. B. 1853c. On the insect beds in the Purbeck Formation of Dorset and Wilts; and a notice of the occurrence of a Neuropterous Insect in the Stonesfield Slate of Gloucestershire. *Quarterly Journal of the Geological Society*, 9(1–2), 344–344.

Coram, R. A., and Jepson, J. E. 2012. *Fossil Insects of the Purbeck Limestone Group of southern England: Palaeoentomology from the Dawn of the Cretaceous* (Vol. 3). Siri Scientific Press, Manchester.

Westwood, J. O. 1854. Contributions to fossil entomology. *Quarterly Journal of the Geological Society*, 10(1–2), 378–396.

Woodward, H. B. 1897. Eminent living geologists: The Rev. P. B. Brodie, MA FGS. *Geological Magazine*, 4, 480–485.

Woodward, H. B. 1897. The Rev. P. B. Brodie, M. A., F. G. S. *Nature*, 57, 31–32.

Buckland, Mary

Conybeare, W. D., Buckland, W., and Buckland, M. 1840. *Ten Plates Comprising a Plan, Sections and Views Representing the Changes Produced on the Coast ff East Devon Between Axmouth and Lyme Regis by the Subsidence of the Land and the Elevation of the Bottom of the Sea, on 25th December 1839 and 3rd February 1840*. John Murray, London.

Cuvier, R. 1817. *Le Règne Animal Distribué d'après son Organisation: Les Mammifères et les Oiseaux* (Vol. 1). Deterville, Paris.

Kölbl-Ebert, M. 1997. Mary Buckland (née Morland) 1797–1857. *Earth Sciences History*, 16, 33–38.

Morgan, N. 2019. Distant thunder. Behind every good man . . . . *Geoscientist*, 29(6), 26.

Oldroyd, D., and Kölbl-Ebert, M. 2012. Sketching rocks and landscape: Drawing as a female accomplishment in the service of Geology. *Earth Sciences History*, 31, 270–286.

Torrens, H. S. 2004. Buckland [*née* Morland], Mary (1797–1857). *Oxford Dictionary of National Biography*, https://doi.org/10.1093/ref:odnb/46486.

Buckland, William

Annan, N. 1999. *The Dons: Mentors, Eccentrics and Geniuses*. University of Chicago Press, Chicago and Harper Collins, London.

Buckland, W. 1822. On the excavation of valleys by diluvian action, as illustrated by a succession of valleys which intersect the South Coast of Dorset and Devon. *Transactions of the Geological Society of London*, 2, 95–102.

Buckland, W. 1823. *Reliquiae Diluvianae, or, Observations on the Organic Remains attesting the Action of a Universal Deluge*. John Murray, London.

Buckland, W. 1824, Notice on the *Megalosaurus* or great Fossil Lizard of Stonesfield, *Transactions of the Geological Society of London*, 2, 390–396.

Buckland, W. 1828. On the Cycadeoideæ, a family of fossil plants found in the Oolite Quarries of the Isle of Portland. *Transactions of the Geological Society of London*, 2(3), 395–401.

Buckland, W. 1829a. On the discovery of a new species of pterodactyle in the Lias at Lyme Regis. *Transactions of the Geological Society of London*, 2(1), 217–222.

Buckland, W. 1829b. On the discovery of coprolites, or fossil fæces, in the Lias at Lyme Regis, and in other formations. *Transactions of the Geological Society of London*, 2(1), 223–236.

Buckland, W. 1835. On the discovery of fossil bones of the Iguanodon, in the Iron Sand of the Wealden Formation in the Isle of Wight, and in the Isle of Purbeck. *Transactions of the Geological Society of London*, 2(3), 425–432.

Buckland, W. 1837. *Geology and Mineralogy Considered with Reference to Natural Theology* (2 Vols.). William Pickering, London.

Buckland, W., and De La Beche, H. T. 1835. On the geology of the neighbourhood of Weymouth and the adjacent parts of the coast of Dorset. *Transactions of the Geological Society of London*, 2(1), 1–46.

Chapman, S. 2020. *Caves, Coprolites and Catastrophes. The Story of Pioneering Geologist and Fossil-Hunter*. SPCK, London.
Conybeare, W. D., Buckland, W., and Buckland, M. 1840. *Ten Plates Comprising a Plan, Sections and Views Representing the Changes Produced on the Coast of East Devon Between Axmouth and Lyme Regis by the Subsidence of the Land and the Elevation of the Bottom of the Sea, on 25th December 1839 and 3rd February 1840*. John Murray, London.
Duffin, C. J. 2006. William Buckland (1784–1856). *Geology Today*, 22(3), 104–108.
Duffin, C. J. 2010. Coprolites. In A. R. Lord and P. G. Davis (eds.), *Fossils from the Lower Lias of the Dorset Coast*. Palaeontological Association, London, 395–400.
Rudwick, M. J. S. 2008. *Worlds Before Adam. The Reconstruction of Geohistory in the Age of Reform*. University of Chicago Press, Chicago and London.
Rupke, N. A. 1983. *The Great Chain of History: William Buckland and the English School of Geology (1814–1849)*. Oxford University Press, Oxford.

Buckman

Anon. 1929. Mr. S. S. Buckman. *Nature*, 123, 419.
Buckman, J. 1873. On the cephalopoda-bed and the oolite sands of Dorset and part of Somerset. *Quarterly Journal of the Geological Society*, 29(1–2), 504–505.
Buckman, S. S. 1878. On the species of Astarte from the Inferior Oolite of the Sherborne District. *Proceedings of the Dorset Natural History and Antiquarian Field Club*, 2, 81–92.
Buckman, S. S. 1881. A descriptive catalogue of some of the species of ammonites from the Inferior Oolite of Dorset. *Quarterly Journal of the Geological Society*, 37, 588–608.
Buckman, S. S. 1910. Certain Jurassic (Lias-Oolite) strata of south Dorset; and their correlation. *Quarterly Journal of the Geological Society*, 66, 52–89.
Buckman, S. S. 1910. *Mating, Marriage and the Status of Woman*. Walter Scott Publishing Co., London and Felling-on-Tyne.
Callomon, J. H. 1995. Time from fossils: S. S. Buckman and Jurassic high-resolution geochronology. *Geological Society, London, Memoirs*, 16(1), 127–150.
Morley-Davies, A. 1930. The geological life-work of Sydney Savory Buckman. *Proceedings of the Geologists' Association*, 41, 221–240.
Torrens, H. S. 2004. Buckman, Sydney Savory (1860–1929). *Oxford Dictionary of National Biography*, https://doi.org/10.1093/ref:odnb/54379.
Torrens, H. S. 2009. James Buckman (1814–1884): The scientific career of an English Darwinian thwarted by religious prejudice. *Geological Society, London, Special Publications*, 310, 245–258.

Burton

Anon. https://en.wikipedia.org/wiki/Ernest_St._John_Burton.
Burton, E. S. J. 1932. A peneplain and re-excavated valley floors in Dorsetshire. *Geological Magazine*, 69(10), 474–477.
Burton, E. S. J. 1937. The origin of Lulworth Cove, Dorsetshire. *Geological Magazine*, 74(8), 377–383.

Carr

Bray, M. J., Duane, W. J. 2005. *GPS Survey of Control Markers on Chesil Beach*. Technical Paper, Dept. Geography, University of Portsmouth, Portsmouth.

Carr, A. P. 1969. Size grading along a pebble beach; Chesil Beach, England. *Journal of Sedimentary Research*, 39(1), 297–311.

Carr, A. P. 1971. Experiments on longshore transport and sorting of pebbles; Chesil Beach, England. *Journal of Sedimentary Research*, 41(4), 1084–1104.

Carr, A. P. 1980. *Chesil Sea Defence Scheme: Assessment of Environmental Implications.* Unpublished Report to the Wessex Water Authority, Bristol.

Carr, A. P. 1980. *Chesil Beach and Adjacent Area: Outline of Existing Data and Suggestions for Future Research.* Report to the Dorset County Council and Wessex Water Authority. Institute of Oceanographic, Wormley.

Carr, A. P. 1983. Chesil Beach: Environmental, economic and sociological pressures. *Geographical Journal*, 149(1), 53–62.

Carr, A. P. 2000. Chesil Beach: Recent changes in a longer-term context. In A. P. Carr, D. R. Seaward, and P. H. Sterling (eds.), *The Fleet Lagoon and Chesil Beach: Proceedings of the Third Symposium of the Fleet Study Group* (Revised edn.), 23–30. Dorset County Council, Dorchester.

Carr, A. P., and Blackley, M. W. L. 1969. Geological composition of the pebbles of Chesil Beach, Dorset. *Proceedings of the Dorset Natural History and Archaeological Society*, 90, 133–140.

Carr, A. P., and Blackley, M. W. L. 1973. Investigations bearing on the age and development of Chesil Beach, Dorset, and the associated area. *Transactions of the Institute of British Geographers*, 58, 99–111.

Carr, A. P., and Blackley, M. W. L. 1974. Ideas on the origin and development of Chesil Beach, Dorset. *Proceedings of the Dorset Natural History and Archaeological Society*, 95, 9–17.

Carr, A. P., and Gleason, R. 1972. Chesil Beach, Dorset and the cartographic evidence of Sir John Coode. *Proceedings of the Dorset Natural History and Archaeological Society*, 93, 125–131.

Carr, A. P., Gleason, R., and King, A. 1970. Significance of pebble size and shape in sorting by waves. *Sedimentary Geology*, 4(1–2), 89–101.

Carr, A. P., and Seaward, D. R. 1990. Chesil Beach: Changes in crest height 1969–1990. *Proceedings of the Dorset Natural History and Archaeological Society*, 112, 109–112.

Carr, A. P., and Seaward, D. R. 1991. Chesil Beach-landward recession 1965–1991. *Proceedings of the Dorset Natural History and Archaeological Society*, 113, 157–160.

Coombe, E. D. K. 1998. *Holocene Palaeoenvironments of the Fleet Lagoon.* PhD Thesis, University of Oxford.

Coombe, E. D. K. 2000. Cored material from the Fleet: Some initial inferences. In A. Carr, D. Seaward, and R. Sterling (eds.), *The Fleet Lagoon and Chesil Beach.* The Fleet Study Group 3rd Symposium Proceedings, 31–40. Dorset County Council, Dorchester.

Whittaker, J. E. 2003. Alan Paul Carr 1930–2000, *Proceedings of the Dorset Natural History and Archaeological Society*, 123, 153–156.

Conybeare

Conybeare, W. D. 1824. On the discovery of an almost perfect skeleton of the Plesiosaurus. *Transactions of the Geological Society of London*, 2(2), 381–338.

Conybeare, W. D., Buckland, W., and Buckland, M. 1840. *Ten Plates Comprising a Plan, Sections and Views Representing the Changes Produced on the Coast of East Devon Between Axmouth and Lyme Regis by the Subsidence of the Land and the Elevation of the Bottom of the Sea, on 25th December 1839 and 3rd February 1840.* John Murray, London.

Conybeare, W. D., and Phillips, W. 1822. *The Geology of England and Wales*. Phillips, London.

De la Beche, H. T., and Conybeare, W. D. 1821. Notice of the discovery of a new fossil animal, forming a link between the Ichthyosaurus and Crocodile, together with general remarks on the osteology of the Ichthyosaurus. *Transactions of the Geological Society of London*, 1(1), 559–594.

Gallois, R. W. 2010. The failure mechanism of the 1839 Bindon Landslide, Devon, UK: Almost right first time. *Geoscience in South-West England: Proceedings of the Ussher Society*, 12, 188–197.

Portlock, G. 1859. Biographical notice of Dean Conybeare and Alcide D'Orbigny. *American Journal of Science and Arts*, 27(79), 63–77.

Torrens, H. S. 2004. Conybeare, William Daniel (1787–1857). *Oxford Dictionary of National Biography*, https://doi.org/10.1093/ref:odnb/6129.

Coade

Freestone, I. 1991. Forgotten but not lost: The secret of Coade Stone. *Proceedings of the Geologists' Association*, 102(2), 135–138.

Kelly, A. 1990. *Mrs Coade's Stone*. Self-Publishing Association, Upton-upon-Severn.

Lemmen, H. van, 2006. *Coade Stone*. Shire Publications, Oxford.

Coode

Anon. 1893. Obituary. Sir John Coode, KCMG, 1816–1892. (President 1889–1891). *Minutes of the Proceedings of the Institution of Civil Engineers*, 113, 334–334.

Beare, T. H., revised by Alan Muir Wood. 2004. Coode, Sir John (1816–1892). *Oxford Dictionary of National Biography*, https://doi.org/10.1093/ref:odnb/6133.

Carr, A. P., and Gleason, R. 1972. Chesil Beach, Dorset and the cartographic evidence of Sir John Coode. *Proceedings of the Dorset Natural History and Archaeological Society*, 93, 125–131.

Coode, J. 1853. Description of the Chesil Bank, with remarks upon its origin, the causes which have contributed to its formation and upon the movement of shingle generally (including appendices and plate). *Minutes of the Proceedings of the Institution of Civil Engineers*, 12, 520–546.

Prestwich, J. 1875. On the origin of Chesil Bank, and on the relation of the existing beaches to past geological changes independent of the present coast action. *Minutes of the Proceedings of the Institution of Civil Engineers*, 40, 61–79.

Reade, T. M., Topley, W., Fisher, O., et al. 1875. Discussion on the origin of Chesil Bank and on the relation of the existing beaches to past geological changes independent of the present coast action. *Minutes of the Proceedings of the Institution of Civil Engineers*, 40, 80–114.

Cornish

Anon. 1948. Dr. Vaughan Cornish. *Nature*, 161, 839.

Cornish, C. J. 1895. *Wild England of Today*. Seeley and Co., London.

Cornish, V. 1898. On sea-beaches and sandbanks. *Geographical Journal*, 11(6), 68–109, 113–121.

Cornish, V. 1930. *National Parks and the Heritage of Scenery*. Sifton Praed, London.

Cornish, V. 1932. *The Scenery of England*. Council for the Preservation of Rural England, London.

Cornish, V. 1937. *The Preservation of Our Scenery.* Cambridge, Cambridge University Press.

Cornish, V. 1940. *The Scenery of Sidmouth.* Cambridge University Press, Cambridge.

Cornish, V. 1942. *A Family of Devon, Their Homes, Travels and Occupations.* St Leonards, King Bros., and Potts, London.

Crone, G. R., and Matless, D. 2004. Cornish, Vaughan (1862–1948). *Oxford Dictionary of National Biography,* https://doi.org/10.1093/ref:odnb/32573.

Goudie, A. 1972. Vaughan Cornish: Geographer (with a bibliography of his published works). *Transactions of the Institute of British Geographers,* 55, 1–16.

Damon

Damon, R. 1860. *Handbook to the Geology of Weymouth and the Island of Portland: With Notes on the Natural History of the Const. & Neighbourhood. Accompanied by a. Map of the District, Geological Sections, Plates of Fossils, Coast Views, and numerous other Illustrations.* Stanford, London.

Damon, R. 1884. *Geology of Weymouth, Portland, and coast of Dorsetshire, from Swanage to Bridport-on-the-sea: With Natural History and Archæological Notes.* R. F. Damon, Weymouth.

Torrens, H. 2004. How the long history of geological studies in Dorset confirms its World Heritage Coast status. *Open University Geological Society Journal,* 25(2), 1–16.

Day

Day, E. C. H. 1863. On the Middle and Upper Lias of the Dorsetshire coast. *Quarterly Journal of the Geological Society,* 19(1–2), 278–297.

Day, E. C. H. 1865. On the Lower Lias of Lyme Regis. *Geological Magazine,* 2, 518–519.

De la Beche

Bate, D. G. 2010. Sir Henry Thomas De la Beche and the founding of the British Geological Survey. *Mercian Geologist,* 17(3), 149–165.

Buckland, W., and De La Beche, H. T. 1835. On the Geology of the neighbourhood of Weymouth and the adjacent parts of the Coast of Dorset. *Transactions of the Geological Society of London,* 2(1), 1–46.

Clary, R., and Wandersee, J. 2014. The journey from elite society to government geologist: Henry De La Beche's (1796–1855) powerful impact on the importance of observation within an emerging professional science. *Earth Sciences History,* 33(2), 259–278.

Clary, R. M., and Wandersee, J. H. 2009. All are worthy to know the Earth: Henry De la Beche and the origin of geological literacy. *Science & Education,* 18(10), 1359–1375.

De la Beche, H. 1839. *Report on the Geology of Cornwall, Devon and West Somerset.* Longman, London.

De la Beche, H. T. 1822. Remarks on the geology of the south coast of England, from Bridport Harbour, Dorset, to Babbacombe Bay, Devon. *Transactions of the Geological Society of London,* 2(1), 40–47.

De la Beche, H. T. 1826. On the Lias of the coast, in the vicinity of Lyme Regis, Dorset. *Transactions of the Geological Society of London,* 2(1), 21–30.

De la Beche, H. T., and Conybeare, W. D. 1821. Notice of the discovery of a new fossil animal, forming a link between the Ichthyosaurus and Crocodile, together with general remarks on the osteology of the Ichthyosaurus. *Transactions of the Geological Society of London,* 1(1), 559–594.

Lam, C. 2021. Decolonising geoscience. *Geoscientist,* Spring 2021, 37–39.

McCartney, P. J. 1977. *Henry De la Beche: Observations of an Observer*. Friends of the National Museum of Wales, Cardiff.

Morgan, N. 2012. https://www.geolsoc.org.uk/Geoscientist/Archive/February-2014/Distant-Thunder-Ladies-Man.

Morris, R. (ed.), 2013. *A Journal of Sir Henry De la Beche: Pioneer Geologist (1796–1855): Written in His Own Hand*. Royal Institution of South Wales, Cardiff.

Norman, D. B. 2000. Henry De la Beche and the plesiosaur's neck. *Archives of Natural History*, 27(1), 137–148.

Reyment, R. A. 1996. Henry Thomas De la Beche (1796–1855). *Terra Nova*, 8, 489–492.

Rudwick, M. J. S. 1992. *Scenes from Deep Time. Early Pictorial Representations of the Prehistoric World*. University of Chicago Press, Chicago and London.

Secord, J. A. 2004. Beche, Sir Henry Thomas De la (1796–1855). *Oxford Dictionary of National Biography*, https://doi.org/10.1093/ref:odnb/1891.

Shaw, K. 2016. http://www.lymeregismuseum.co.uk/lrm/wp-content/uploads/2016/08/sir_henry_de_la_beche_in_lyme_regis.pdf.

Fisher

Davison, C. 1900. Eminent living geologists: Rev. Osmond Fisher, MA, FGS. *Geological Magazine*, 7(2), 49–54.

Fisher, O. 1851. On the Purbeck strata of Dorsetshire. *Transactions of the Cambridge Philosophical Society*, 9, 555–581.

Fisher, O. 1859. On some natural pits on the heaths of Dorsetshire. *Quarterly Journal of the Geological Society*, 15(1–2), 187–188.

Fisher, O. 1873. On the origin of the estuary of the Fleet in Dorsetshire. *Geological Magazine*, 10(113), 481–482.

Fisher, O. 1881. *Physics of the Earth's Crust*. Macmillan, London.

Fisher, O. 1888. On the occurrence of Elephas meridionalis at Dewlish, Dorset. *Quarterly Journal of the Geological Society*, 44(1–4), 818–824.

Fisher, O. 1896. Vertical tertiaries at Bincombe, Dorset. *Geological Magazine*, 3(6), 246–247.

Fisher, O. 1905. On the occurrence of Elephas meridionalis at Dewlish (Dorset). Second communication: Human agency suggested. *Quarterly Journal of the Geological Society*, 61(1–4), 35–38.

Wilding, R. 1988. Osmond Fisher (1817–1914). *Proceedings of the Dorset Natural History and Archaeological Society*, 110, 17–22.

Wilding, R. 2004. Fisher, Osmond (1817–1914). *Oxford Dictionary of National Biography*, https://doi.org/10.1093/ref:odnb/56925.

Fitton

Crabbe, R. M. n.d. The inspiring story of William Henry Fitton. www.irishancestors.ie/william-henry-fitton-1780-1861.

Eyles, J. 1961. Dr. W. H. Fitton, F. R. S. *Nature* 190, 585.

Fitton, W. H. 1836, *Observations on some of the Strata between the Chalk and the Oxford Oolite in the Southeast of England*. R. Taylor, London.

Manning, P. I., and Manning, J. M. E. 1961. The centenary of a pioneer Irish geologist: W. H. Fitton, MD, FRS, 1780–1861. *The Irish Naturalists' Journal*, 13, 241–244.

Murchison, R. I. 1862. Obituary notice of Dr Fitton. *Proceedings of the Geological Society*, 18, xxx–xxxiv.

Torrens, H. S., and Browne, J. 2004. Fitton, William Henry (1780–1861). *Dictionary of National Biography*, https://doi.org/10.1093/ref:odnb/9525.

Forbes

Balfour, J. H. 1858. Sketch of the life of the late Professor Edward Forbes. *Transactions of the Botanical Society of Edinburgh*, 5, 23–41.

Egerton, F. N. 2010. History of the ecological sciences, Part 35: The beginnings of British Marine Biology: Edward Forbes and Philip Gosse. *Bulletin of the Ecological Society of America*, 91(2), 176–201.

Forbes, E. 1850. On the succession of strata and distribution of organic remains in the Dorsetshire Purbecks. *Report of the British Association for 1850*, 79–81.

Gardiner, B. G. 1993. Edward Forbes, Richard Owen and the Red Lions. *Archives of Natural History*, 20(3), 349–372.

Huxley, T. 1854. Professor Edward Forbes, FRS. *Journal of Science and Literary Gazette*, 1854, 1016–1018.

Merriman, D. 1963. Edward Forbes: Manxman. *Progress in Oceanography*, 3, 191–206.

Mills, E. L. 2004. Forbes, Edward (1815–1854). *Oxford Dictionary of National Biography*, https://doi.org/10.1093/ref:odnb/9824.

Preece, R. C. 1995. Edward Forbes (1815–1854) and Clement Reid (1853–1916): Two generations of pioneering polymaths. *Archives of Natural History*, 22(3), 419–435.

Rehbock, P. F. 1979. Edward Forbes (1815–1854): An annotated list of published and unpublished writings. *Journal of the Society for the Bibliography of Natural History*, 9(2), 171–218.

Rehbock, P. F. 2001. Edward Forbes. *Wiley Encyclopedia of Life Sciences*, https://onlinelibrary.wiley.com/doi/pdf/10.1038/npg.els.0002518.

Ritchie, J. 1956. A double centenary: Two notable naturalists, Robert Jameson and Edward Forbes. *Proceedings of the Royal Society of Edinburgh, Section B: Biological Sciences*, 66(1), 29–58.

Wilson, G., and Geikie, A. 1861. *Memoir of Edward Forbes, Late Regius Professor of Natural History in the University of Edinburgh*. MacMillan, London.

Green

E. E. S. B. 1949. Obituary notice. *Quarterly Journal of the Geological Society*, 105, lxvi–lxix.

Green, J. F. N. 1936. The terraces of southernmost England. *Quarterly Journal of the Geological Society*, 92(2), 58–88.

Green, J. F. N. 1941. The high platforms of east Devon. *Proceedings of the Geologists' Association*, 52(1), 36–52.

Green, J. F. N. 1943. The age of the raised beaches of South Britain. *Proceedings of the Geologists' Association*, 54(3), 129–140.

Green, J. F. N. 1947. Some gravels and gravel-pits in Hampshire and Dorset. *Proceedings of the Geologists' Association*, 58(2), 128–143.

Hawkins

Carroll, V. 2007. 'Beyond the pale of ordinary criticism'. Eccentricity and the fossil books of Thomas Hawkins. *Isis*, 98(2), 225–265.

Hawkins, T. 1834. *Memoirs on Ichthyosauri and Plesiosauri*. Relfe and Fletcher, London.

Hawkins, T. 1840. *The Great Sea Dragons*. William Pickering, London.

McGowan, C. 2001. *The Dragon Seekers: The Discovery of Dinosaurs During the Prelude to Darwin*. Little Brown, London.

Norman, D. B. 2000. Professor Richard Owen and the important but neglected dinosaur *Scelidosaurus harrisonii*. *Historical Biology*, 14(4), 235–253.

O'Connor, R. 2003. Thomas Hawkins and geological spectacle. *Proceedings of the Geologists' Association*, 114(3), 227–241.

Purcell, R. W., and Gould, S. J. 1993. *Finders, Keepers, Eight Collectors*. Pimlico, London.

Taylor, M. A. 2003. Joseph Clark III's reminiscences about the Somerset fossil reptile collector Thomas Hawkins (1810–1889): 'Very near the borderline between eccentricity and criminal insanity'. *Proceedings of the Somerset Archaeological and Natural History Society*, 146, 1–10.

Taylor, M. A. 2004. Hawkins, Thomas (1810–1889). *Oxford Dictionary of National Biography*, https://doi.org/10.1093/ref:odnb/12682.

Hobbs

Hobbs, W. *c*. 1715. *The Earth Generated and Anatomized.*

Porter, R. 1976. William Hobbs of Weymouth and his *The Earth Generated and Anatomized* (? 1715). *Journal of the Society for the Bibliography of Natural History*, 7(4), 333–341.

Porter, R. (ed.). 1981. *The Earth Generated and Anatomized by William Hobbs: An Early Eighteenth Century Theory of the Earth*. British Museum, London.

Hooke

Chapman, A. 2004. *England's Leonardo: Robert Hooke and the Seventeenth-century Scientific Revolution*. Institute of Physics, Bristol and Philadelphia.

Davies, G. L. 1964. Robert Hooke and his conception of earth-history. *Proceedings of the Geologists' Association*, 75(4), 493–498.

Davies, G. L. 1969. *The Earth in Decay*. Macdonald, London.

Drake, E. T. 1996. *Restless Genius: Robert Hooke and His Earthly Thoughts*. Oxford University Press, New York.

Drake, E. T., Komar, P. D. 1981. A comparison of the geological contributions of Nicolaus Steno and Robert Hooke. *Journal of Geological Education*, 29(3), 127–134.

Drake, E. T., and Komar, P. D. 1983. Speculations about the Earth: The role of Robert Hooke and others in the 17th century. *Earth Sciences History* 2, 11–16.

Hooke, R. 1665. *Micrographia, or some Physiological Descriptions of Minute Bodies made by Magnifying Glasses, with Observations and Inquiries Thereupon*. Martyn, London.

Hooke, R. 1705. *Lectures and Discourses of Earthquakes and Subterraneous Eruptions*. Waller, London.

Oldroyd, D. R. 1972. Robert Hooke's methodology of science as exemplified in his 'Discourse of Earthquakes'. *British Journal for the History of Science*, 6(2), 109–130.

House

Becker, R. T., and Kirchgasser, W. T. 2007. Devonian events and correlations: A tribute to the lifetime achievements of Michael Robert House (1930–2002). *Geological Society, London, Special Publications*, 278(1), 1–8.

Butcher, N. E. 2002. Obituary of Michael House. https://www.geolsoc.org.uk/en/About/History/Obituaries%202001%20onwards/Obituaries%202002/Michael%20Robert%20House%201930-2002.

House, M. R. 1955. New records from the Red Nodule Beds near Weymouth. *Proceedings of the Dorset Natural History and Archaeology Society*, 75, 134–135.

House, M. R. 1958. *Geologists' Association Guides No. 22: The Dorset Coast from Poole to Chesil Beach*. Geologists' Association, London.

House, M. R. 1961. The structure of the Weymouth Anticline. *Proceedings of the Geologists' Association*, 72(2), 221–223.

House, M. R. 1968. Purbeckian calcareous algae. *Proceedings of the Dorset Natural History and Archaeological Society*, 89, 42–45.

House, M. R. 1969. Portland Stone on Portland. *Proceedings of the Dorset Natural History and Archaeological Society*, 91, 38–39.

House, M. R. 1986. Are Jurassic sedimentary microrhythms due to orbital forcing? *Proceedings of the Ussher Society*, 6(3), 299–311.

House, M. R. 1989. *Geology of the Dorset Coast (No. 22)*. Geologists' Association, London.

House, M. R. 1991. Dorset dolines: Part 1, The higher Kingston road cutting. *Proceedings of the Dorset Natural History and Archaeological Society*, 112, 105–108.

House, M. R. 1992. Dorset dolines: Part 2, Bronkham Hill. *Proceedings of the Dorset Natural History and Archaeological Society*, 113, 149–155.

House, M. R. 1995. Dorset dolines: Part 3. Eocene pockets and gravel pipes in the chalk of St Oswald's Bay. *Proceedings of the Dorset Natural History and Archaeological Society*, 117, 109–116.

Neale, J. W. 2002. Michael Robert House (1930–2002). *Proceedings of the Yorkshire Geological Society*, 54, 127.

Hutchinson

East Devon Area of Outstanding Natural Beauty. 2012. http://www.eastdevonaonb.org.uk/our-work/culture-and-heritage/poh.

Hutchinson, P. O. (writing as P. O. H. Sidmouthiensis). 1840. *A Guide to the Landslip, near Axmouth, Devonshire: Together with a Geological and Philosophical Enquiry into its Nature and Causes, and a Topographical Description of the District. Illustrated by Engravings*. John Harvey, Sidmouth.

Hutchinson, P. O. 1843. *The Geology of Sidmouth and of South-Eastern Devon*. John Harvey, Sidmouth.

Hutchinson, P. O. 1860. *A New Guide to Sidmouth and the Neighbourhood: Comprising Notices of the Towns, Places, and Objects of Interest, Within the Distance of Twenty Miles; Illustrated with Map, Views, and Diagrams*. Thomas Perry, Sidmouth.

Hutchinson, P. O. 1873. Submerged forest and mammoth teeth at Sidmouth. *Transactions of the Devonshire Association for the Advancement of Science, Literature and Art*, 6, 232–235.

Hutchinson, P. O. 1886. Sidmouth: Rate of erosion of the sea-coasts of England and Wales. *Report of the 1885 Meeting of the British Association for the Advancement of Science*, 417–422.

Hutchinson, P. 1893. Landslip at Sidmouth. *Reports and Transactions of the Devonshire Association*, 25, 174–175.

Hutchinson, P. 1906. Geological section of the cliffs to the west and east of Sidmouth, Devon. *Report of the British Association for the Advancement of Science*, 168–170.

Mather, J. D., and Symes, R. F. 2006. Peter Orlando Hutchinson (1810–1897). *Geoscience in South-west England*, 11, 214–221.

Jukes-Browne

J. W. J. 1914. Alfred John Jukes-Browne, F. R. S. *Nature*, 93, 67–68.

Jukes-Browne, A. J. 1892. *The Building of the British Isles* (2nd edn.). George Bell, London.

Jukes-Browne, A. J. 1898. The Origin of the Vale of Marshwood in West Dorset. *Geological Magazine*, 5(4), 161–168.

Jukes-Browne, A. J. 1902. On a deep boring at Lyme Regis. *Quarterly Journal of the Geological Society*, 58(1–4), 279–289.

Jukes-Browne, A. J. 1908. The burning cliff and the landslide at Lyme Regis. *Proceedings of Dorset Natural History and Antiquarian Field Club*, 29, 153–160.

Woodward, H. B., Ussher, W. A. E., and Jukes-Browne, A. J. 1911. *The Geology of the Country near Sidmouth and Lyme Regis* (Vol. 326). H. M. S.O, London.

Lang

Hallam, A. 1989. W. D. Lang's research on the Lias of Dorset. *Proceedings of the Geologists' Association*, 100(4), 451–455.

Lang, W. D. 1903. On a fossiliferous bed in the Selbornian of Charmouth. *Geological Magazine*, 9, 388–392.

Lang, W. D. 1907. The Selbornian of Stonebarrow Cliff, Charmouth. *Geological Magazine*, 4(4), 150–156.

Lang, W. D. 1913. The Lower Pliensbachian—'Carixian'—of Charmouth. *Geological Magazine*, 10(9), 401–412.

Lang, W. D. 1914. The geology of the Charmouth cliffs, beach and fore-shore. *Proceedings of the Geologists' Association*, 25(5), 293–362.

Lang, W. D. 1926. The submerged forest at the mouth of the River Char and the history of that river. *Proceedings of the Geologists' Association*, 37(2), 197–210.

Lang, W. D. 1932. The Lower Lias of Charmouth and the Vale of Marshwood. *Proceedings of the Geologists' Association*, 43(2), 97–126.

Lang, W. D. 1936. The green ammonite beds of the Dorset Lias. *Quarterly Journal of the Geological Society*, 92, 423–437.

Lang, W. D. 1955. Mudflows at Charmouth. *Proceedings of the Dorset Natural History and Archaeological Society*, 75, 151–156.

Lang, W. D., Spath, L. F., Cox, L. R., and Muir-Wood, H. M. 1928. The Belemnite marls of Charmouth, a series in the Lias of the Dorset Coast. *Quarterly Journal of the Geological Society*, 84(1–4), 179–222.

Lang, W. D., Spath, L. F., and Richardson, W. A. 1923. Shales-with-'beef,' a Sequence in the Lower Lias of the Dorset Coast. *Quarterly Journal of the Geological Society*, 79(1–4), 47–66.

White, E. I. 1966. William Dickson Lang 1878–1966. *Biographical Memoir of the Royal Society*, 12, 366–386.

Mansel-Pleydell

Anon. 1902. John Clavell Mansel-Pleydell, F. L. S., F. G. S. *Geological Magazine*, 9, 335–6.

Fryer, S. E. 1912. Mansel-Pleydell, John Clavell (1817–1902). *Oxford Dictionary of National Biography*, https://doi.org/10.1093/odnb/9780192683120.013.34860.

Mansel-Pleydell, J. C. 1886. On a tufaceous deposit at Blashenwell, Isle of Purbeck. *Proceedings of the Dorset Field Club*, 1, 109–113.

Mansel-Pleydell, J. C. 1874/1895. *The Flora of Dorsetshire: With a Sketch of the Topography, River System, and Geology of the County*. Dorset County Chronicle Printing Works, Dorchester.

Mansel-Pleydell, J. C. 1888. *The Birds of Dorsetshire: A Contribution to the Natural History of the County*. RH Porter, London.

Mansel-Pleydell, J. C. 1896. On the footprints of a dinosaur (Iguanodon?) from the Purbeck Beds of Swanage. *Proceedings of the Dorset Natural History and Antiquarian Field Club*, 18, 115.

Mansel-Pleydell, J. C. 1898. *The Mollusca of Dorsetshire*. Privately printed.

Wills, S. 2013. John Whitaker Hulke, surgeon and palaeontologist. *Geological Society, London, Special Publications*, 375(1), 409–427.

Owen

Gaudant, J. 1992. Hommage à Richard Owen (1804–1892), le Cuvier Anglais. *Travaux du Comité français d'Histoire de la Géologie*, 3 (tome 6), 109–117.

Gruber, J. W. 2004. Owen, Sir Richard (1804–1892). *Oxford Dictionary of National Biography*,https://doi.org/10.1093/ref:odnb/21026.

Norman, D. B. 2000. Professor Richard Owen and the important but neglected dinosaur Scelidosaurus harrisonii. *Historical Biology*, 14(4), 235–253.

Owen, R. 1841. *Report on British Fossil Reptiles*. R. and J. E. Taylor, London.

Owen, S. R. 1862. *A Monograph of a Fossil Dinosaur, Scelidosaurus harrisonii, Owen, of the Lower Lias*. Palaeontographical Society, London.

Padian, K. 1997. The rehabilitation of Sir Richard Owen. *BioScience*, 47(7), 446–453.

Rupke, N. A. 2009. *Richard Owen: Biology without Darwin*. University of Chicago Press, Chicago.

Parkinson

Compston, D. A. S. 1991. Book review. James Parkinson: His life and times by A. D. Morris. *Journal of Neurology*, 238(3), 129–130.

Lewis, C. 2017a. The Enlightened Mr. Parkinson. *Geoscientist*, 27(3), 17–19.

Lewis, C. 2017b. *The Enlightened Mr. Parkinson: The Pioneering Life of a Forgotten English Surgeon*. Icon Books, London.

Lewis, C. L. 2009. 'Our favourite science': Lord Bute and James Parkinson searching for a Theory of the Earth. *Geological Society, London, Special Publications*, 310(1), 111–126.

Liston, J. J., Alcalá, L. 2017. The obstetrician, the surgeon and the premature birth of the world's first dinosaur: William Hunter and James Parkinson. *Geological Society, London, Special Publications*, 452(1), 77–101.

Parkinson, J. 1804–1811. *Organic Remains of a Former World. An examination of the Mineralized Remains of the Vegetables and Animals of the Antediluvian World, Generally Termed Extraneous Fossils* (published in three volumes). Robson, London.

Rose, F. C. 2013. *James Parkinson: His Life and Times*. Birkhauser, Boston.

Thackray, J. C. 1976. James Parkinson's 'Organic Remains of a Former World' (1804–1811). *Journal of the Society for the Bibliography of Natural History*, 7(4), 451–466.

Philpot

Agassiz, L. 1843. *Recherches sur les Poissons Fossiles*. Petitpierre, Neuchâtel, Switzerland.

Buckland, W. 1829. On the discovery of a new species of Pterodactyle in the Lias at Lyme Regis. *Transactions of the Geological Society of London*, 3, 217–222.

Davis, L. E. 2012. Mary Anning: Princess of palaeontology and geological lioness. *The Compass: Earth Science Journal of Sigma Gamma Epsilon*, 84(1), 56–88.

Duffin, C. J. 2012. Coprolites and characters in Victorian Britain. Vertebrate Coprolites. *New Mexico Museum of Natural History and Science, Bulletin*, 57, 45–60.Edmonds, J. M. 1978. The fossil collection of the Misses Philpot of Lyme Regis. *Proceedings of the Dorset Natural History & Archaeological Society*, 98, 43–48.

Elder, E. S. 1982. Women in early geology. *Journal of Geological Education*, 30(5), 287–293.
Oldroyd, D., Kolbl-Ebert, M. 2012. Sketching rocks and landscape: Drawing as a female accomplishment in the service of Geology. *Earth Sciences History* 31(2), 270–286.

Prestwich

Gamble, C., and Kruszynski, R. 2009. John Evans, Joseph Prestwich and the stone that shattered the time barrier. *Antiquity*, 83, 461–475.
Prestwich, J. 1875a. Notes on the phenomena of the Quaternary Period in the Isle of Portland and around Weymouth. *Quarterly Journal of the Geological Society*, 31(1–4), 29–54.
Prestwich, J. 1875b. On the origin of Chesil Bank, and on the relation of the existing beaches to past geological changes independent of the present coast action. *Minutes of the Proceedings of the Institution of Civil Engineers*, 40, 61–79.
Prestwich, J. 1892. The raised beaches, and 'head' or rubble-drift, of the South of England: Their relation to the valley drifts and to the Glacial Period; and on a late post-Glacial submergence. *Quarterly Journal of the Geological Society*, 48(1–4), 263–343.
Prestwich, J. 1893. On the evidences of a submergence of Western Europe, and of the Mediterranean coasts, at the close of the glacial or so-called post-glacial period, and immediately preceding the neolithic or recent period. *Philosophical Transactions of the Royal Society of London (A)*, 184, 903–984.
Thackray, J. C. 2004. Prestwich, Sir Joseph (1812–1896). *Oxford Dictionary of National Biography*, https://doi.org/10.1093/ref:odnb/22736.
Woodward, H. B. 1893. Eminent Living Geologists. No. 8. J. Prestwich. *Geological Magazine*, 10(6), 241–246.
Woodward, H. B. 1896. Sir Joseph Prestwich, D. C. L., F. R. S. *Nature*, 54, 202–203.

Reid

Anon. 1916. Clement Reid F. R. S. *Nature*, 98, 312.
Grant, M. 2017. Clement (FRS FLS FGS): Remembering a 19th century pioneer in British Quaternary Science at the centenary of his death. *Quaternary Newsletter* 141, 22–25.
J. E. M., and E. T. N. 1919. Clement Reid, 1853–1916. *Proceedings of the Royal Society, series B*, 90, viii–x.
Preece, R. C. 1995. Edward Forbes (1815–1854) and Clement Reid (1853–1916): Two generations of pioneering polymaths. *Archives of Natural History*, 22(3), 419–435.
Reid, C. 1896. The Eocene deposits of Dorset. *Quarterly Journal of the Geological Society* 52, 490–496.
Reid, C. 1896. An early Neolithic kitchen-midden and tufaceous deposit at Blashenwell, near Corfe Castle. *Dorset Natural History and Antiquarian Field Club*, 17, 67–75.
Reid, C. 1913. *Submerged Forests*. Cambridge University Press, Cambridge.
Shepherd, T. 1922. In Memoriam: Clement Reid. *Proceedings of the Yorkshire Geological Society*, 19, 420–422.

Spath

Cox, L. R. 1957a. Dr. L. F. Spath, F. R. S. *Nature*, 179, 847.
Cox, L. R. 1957b. Leonard Frank Spath 1882–1957. *Biographical Memoirs of the Royal Society*, 3, 217–226.
Lang, W. D., and Spath, L. F. 1926. The black marl of Black Ven and Stonebarrow, in the Lias of the Dorset coast. *Quarterly Journal of the Geological Society*, 82(1–4), 144–165.

Lang, W. D., Spath, L. F., Cox, L. R., and Muir-Wood, H. M. 1928. The Belemnite marls of Charmouth, a series in the Lias of the Dorset coast. *Quarterly Journal of the Geological Society*, 84(1–4), 179–222.

Lang, W. D., Spath, L. F., and Richardson, W. A. 1923. Shales-with-'beef,' a sequence in the Lower Lias of the Dorset Coast. *Quarterly Journal of the Geological Society*, 79(1–4), 47–66.

Spath, L. F. 1920. On a new ammonite genus (Dayiceras) from the Lias of Charmouth. *Geological Magazine*, 57(12), 538–543.

Spath, L. F. 1923. The ammonites of the shales-with-'beef'. *Quarterly Journal of the Geological Society*, 79(1–4), 66–88.

Spath, L. F. 1923. Correlation of the Ibex and Jamesoni zones of the Lower Lias. *Geological Magazine*, 60(1), 6–11.

Spath, L. F. 1924. The ammonites of the Blue Lias. *Proceedings of the Geologists' Association*, 35(3), 186–211.

Spath, L. F. 1936. The ammonites of the green ammonite beds of Dorset. *Quarterly Journal of the Geological Society*, 92(1–4), 438–455.

Spath, L. F., and Wright, C. W. 1982. L. F. Spath (1882–1957), ammonitologist. *Annals of Natural History*, 11, 103–105.

Stopes

Chaloner, W. G. 1959. Obituary: Marie Stopes. *Proceedings of the Geologists' Association*, 70, 118–120.

Chaloner, W. G. 2005. The palaeobotanical work of Marie Stopes. *Geological Society, London, Special Publications*, 241(1), 127–135.

Debenham, C. 2018. *Marie Stopes' Sexual Revolution and the Birth Control Movement*. Springer, Dordrecht.

Falcon-Lang, H. 2008. Marie Stopes: Passionate about palaeobotany. *Geology Today*, 24(4), 132–136.

Hall, L. A. 2004. Stopes [*married name* Roe], Marie Charlotte Carmichael (1880–1958). *Oxford Dictionary of National Biography*, https://doi.org/10.1093/ref:odnb/36323.

Morgan, N. 2019. Distant thunder. Get the picture? *Geoscientist*, 29, 26.

Stopes, M. C. 1918. *Married Love*. Fifield, London.

Stopes, M. C. 1918. *Wise Parenthood: A Treatise on Birth Control or Contraception*. Rendell & Co., London.

Stopes, M. C., Oakley, K. P., and Wells, L. H. 1952. A discovery of human skulls, with stone artefacts and animal bones, in a fissure at Portland. *Proceedings of the Dorset Natural History and Archaeological Society*, 74, 39–47.

Strahan

Anon. 1915. Eminent Living Geologists: A. Strahan. *Geological Magazine*, 2(5), 193–198.

Anon. 1928. Sir Aubrey Strahan K. B. E., F. R. S. *Nature*, 121, 461.

Bate, D. G., and Morrison, A. L. 2018. Some aspects of the British Geological Survey's contribution to the war effort at the Western Front, 1914–1918. *Proceedings of the Geologists' Association*, 129(1), 3–11.

Flett, J. S. 1928. Sir Aubrey Strahan. *Proceedings of the Royal Society of London*, 103, xvi–xx.

Howarth, R. J. 2004. Strahan, Sir Aubrey (1852–1928). *Oxford Dictionary of National Biography*, https://doi.org/10.1093/ref:odnb/30010.

Strahan, A. 1896. On the physical geology of Purbeck. *Proceedings of the Geologists' Association*, 14, 405–408.

Strahan, A. 1898. *The Geology of the Isle of Purbeck and Weymouth*. Geological Survey, London.

Strahan, A. 1901. An abnormal section of Chloritic Marl at Mupe Bay, Dorset. *Geological Magazine*, 8(7), 319–321.

Ussher

Anon. 1920. W. A. E. Ussher. *Nature*, 105, 144.

Burt, E. 2013. W. A. E. Ussher: An insight into his life and career. *Geoscience in South-West England*, 13, 165–171.

Dineley, D. L. 1974. W. A. E. Ussher: His work in the south west. *Proceedings of the Ussher Society*, 3, 189–201.

Ussher, W. A. E. 1875. On the subdivisions of the Triassic Rocks, between the coast of West Somerset and the south coast of Devon. *Geological Magazine*, 2(4), 163–168.

Ussher, W. A. E. 1876. On the Triassic rocks of Somerset and Devon. *Quarterly Journal of the Geological Society*, 32(1–4), 367–394.

Woodward, H. B., Ussher, W. A. E., and Jukes-Browne, A. J. 1911. *The Geology of the Country near Sidmouth and Lyme Regis*. H. M. S.O, London.

Webster

Challinor, J. 1970. The progress of British geology during the early part of the nineteenth century. *Annals of Science*, 26(3), 177–234.

Edwards, N. 1971. Thomas Webster (circa 1772–1844). *Journal of the Society for the Bibliography of Natural History*, 5(6), 468–473.

Edwards, N. 2004. Webster, Thomas (1772-1844). *Oxford Dictionary of National Biography*, https://doi.org/10.1093/ref:odnb/28945.

Englefield, H. C., Cooke, G., Cooke, W. B., and Webster, T. 1816. *A Description of the Principal Picturesque Beauties, Antiquities, and Geological Phænomena of the Isle of Wight. With Additional Observations on the Strata of the Island, and Their Continuation in the Adjacent Parts of Dorsetshire, by T. Webster. Illustrated by Maps and Numerous Engravings by W., and G. Cooke, from Original Drawings by Sir HCE, Etc.* Payne and Foss, London.

Heringman, N. 2009. Picturesque ruin and geological antiquity: Thomas Webster and Sir Henry Englefield on the Isle of Wight. *Geological Society, London, Special Publications*, 317(1), 299–318.

Kirk, W. 1996. Thomas Webster (1772–1844): First Professor of Geology at University College London. *Archives of Natural History*, 23(3), 309–326.

Webster, T. 1814. On the fresh-water formations in the Isle of Wight, with some observations on the strata over the Chalk in the South-east part of England. *Transactions of the Geological Society of London*, 1(1), 161–254.

Webster, T. 1826. Observations on the Purbeck and Portland Beds. *Transactions of the Geological Society of London*, 2(1), 37–44.

Woodward, H.

Anon. 1914. H. B. Woodward, F. R. S. *Nature*, 92, 692.

Anon. 1915. Horace Bolingbroke Woodward, F. R. S., F. G. S. *Proceedings of the Geologists' Association*, 26, 142–144.

Evans, M. 2010. The roles played by museums, collections and collectors in the early history of reptile palaeontology. *Geological Society, London, Special Publications*, 343(1), 5–29.

Woodward, H. B. 1876. *The Geology of England and Wales: A Concise Account of the Lithological Characters, Leading Fossils, and Economic Products of the Rocks; with Notes on the Physical Features of the Country*. Longmans, Green and Company, London.
Woodward, H. B. 1892–1895. *The Jurassic Rocks of Britain*. HMSO, London.
Woodward, H. B. 1904. *Stanford's Geological Atlas of Great Britain*. E. Stanford, London.
Woodward, H. B. 1907. *The History of the Geological Society of London*. The Geological Society, London.
Woodward, H. B. 1911. *History of Geology*. Putnam, London.
Woodward, H. B., Ussher, W. A. E., and Jukes-Browne, A. J. 1911. *The Geology of the Country near Sidmouth and Lyme Regis* (Vol. 326). H. M. S.O, London.

Woodward, J.

Challinor, J. 1953. The early progress of British geology. I: From Leland to Woodward, 1538–1728. *Annals of Science*, 9(2), 124–153.
Davies, G. L. 1969. *The Earth in Decay*. McDonald, London.
Delair, J. B., and Sarjeant, W. A. 1975. The earliest discoveries of dinosaurs. *Isis*, 66(1), 5–25.
Eyles, V. A. 1965. John Woodward, F. R. S. (1665–1728), physician and geologist. *Nature*, 206, 868–871.
Eyles, V. A. 1971. John Woodward, FRS, FRCP, MD (1665–1728): A bio-bibliographical account of his life and work. *Journal of the Society for the Bibliography of Natural History*, 5(6), 399–427.
Haycock, D. B. 2002. *William Stukeley*. Boydell Press, Woodbridge.
Levine, J. M. 1991. *Dr. Woodward's Shield: History, Science, and Satire in Augustan England*. Cornell University Press, Ithaca and London.
Levine, J. M. 2004. Woodward, John Woodward, J (1665/1668–1728). *Oxford Dictionary of National Biography*, https://doi.org/10.1093/ref:odnb/29946.
McNamara, K. 2015. Dr Woodward's 350-year legacy. *Geology Today*, 31(5), 181–186.
Porter, R. 1979. John Woodward: 'A droll sort of philosopher'. *Geological Magazine*, 116(5), 335–343.
Price, D. 1989. John Woodward and a surviving British geological collection from the early eighteenth century. *Journal of the History of Collections*, 1(1), 79–95.
Woodward, J. 1695. *An Essay toward a Natural History of the Earth and Terrestrial Bodies, especially Minerals, &c*. Richard Wilkin, London.
Woodward, J. 1696. *Brief Instructions for Making Observations in all Parts of the World*. Richard Wilkin, London.
Woodward, J. 1723. *An Essay towards a Natural History of the Earth: And Terrestrial Bodyes, Especialy Minerals: As Also of the Sea, Rivers, and Springs: With an Account of the Universal Deluge: And of the Effects that it Had Upon the Earth*. A. Bettesworth, London.
Woodward, J. 1728 and 1729. *An Attempt towards a Natural History of the Fossils of England* (2 vols.). Fayram, London.
Woodward, J. 2014. *An Essay towards a Natural History of the Earth*. Cambridge, Cambridge University Press.

Index